T0155500

SpringerBriefs in Applied Sciences and Technology

Manufacturing and Surface Engineering

Series editor

Joao Paulo Davim, Aveiro, Portugal

More information about this series at http://www.springer.com/series/10623

Sumit Bhowmik · Jagadish
Kapil Gupta

Modeling and Optimization of Advanced Manufacturing Processes

 Springer

Sumit Bhowmik
Department of Mechanical Engineering
National Institute of Technology Silchar
Silchar, India

Kapil Gupta
Department of Mechanical and Industrial
  Engineering Technology
University of Johannesburg
Johannesburg, South Africa

Jagadish
Department of Mechanical Engineering
National Institute of Technology Raipur
Raipur, India

ISSN 2191-530X   ISSN 2191-5318 (electronic)
SpringerBriefs in Applied Sciences and Technology
ISSN 2365-8223   ISSN 2365-8231 (electronic)
Manufacturing and Surface Engineering
ISBN 978-3-030-00035-6   ISBN 978-3-030-00036-3   (eBook)
https://doi.org/10.1007/978-3-030-00036-3

Library of Congress Control Number: 2018953644

This Springer imprint is published by the registered company Springer Nature Switzerland AG
The registered company address is: Gewerbestrasse 11, 6330 Cham, Switzerland

Preface

Advanced machining processes are well known to machine difficult-to-machine materials, interacted shapes and features, and micro-parts. Electrical discharge machining (EDM), abrasive water jet machining (AWJM), and ultrasonic machining (USM) are the most extensively used advanced processes. Surface quality, productivity, and sustainability in various advanced processes are mainly dependent on the process parameter combination. In general, it is difficult to manufacture high-quality parts with high productivity and at low environmental footprint only on the basis of operator's skills and referencing machine manual. Therefore, it is essential to research and develop optimum parameter settings as regards to every process for machining any particular material and geometrical shape. Fused deposition modeling (FDM)-type rapid prototyping (RP) method works on the additive layer manufacturing technology have been substituting conventional manufacturing processes. It also necessitates optimal parameter settings to produce quality parts made of plastic and/or metal powders. Moreover, a wide range of input and output process parameters in the aforementioned processes make essential to implement multi-criteria decision-making techniques for optimization.

This book provides the details of the experimentation and modeling and optimization of various advanced manufacturing processes conducted by the authors. It discusses the research results, where EDM, AWJM, USM, and FDM methods are optimized using various multi-criteria decision-making techniques such as subjective + objective weighted method, weighted grey relational analysis, TOPSIS, fuzzy and multi-criteria ranking analysis in order to get optimal material removal rate, surface quality, geometric accuracy, and mechanical properties. Mainly, the effectiveness of the aforementioned multi-criteria decision-making techniques is discussed.

This book consists of five chapters. It starts with Chap. 1 as an introduction to advanced machining processes and previous use of multi-criteria decision-making techniques in there. Chapter 2 presents modeling and optimization of EDM process coupled with an experimental research for optimal values of material removal rate, electrode wear rate, and surface roughness using integrate subjective + objective weighted and grey relational analysis. Chapter 3 reports the effectiveness of decision making and trial evaluation laboratory method integrated with TOPSIS to improve productivity, geometric accuracy, and surface quality in AWJM process. In Chap. 4, integrated fuzzy and multi-criteria ratio analysis has been used to optimize the ultrasonic machining of zirconia ceramic. Finally, optimization of experimental results on the fused deposition modeling based rapid manufacturing of nylon-polyamide parts is presented in Chap. 5.

Authors hope that the research reported on the experimentation, modeling, and optimization of advanced manufacturing would facilitate and motivate the researchers, engineers, and specialists working in this area.

Silchar, India   Jagadish
Silchar, India   Sumit Bhowmik
Johannesburg, South Africa   Kapil Gupta

Contents

Nomenclature

D_{ij}	Decision matrix
Y_{ij}	Performance measure of ith alternatives on jth criteria parameters
m	Number of alternatives
n	Number of criteria
E_{ij}/N_{ij}	Normalized performance weights of ith alternatives on jth criterion
p_j	Entropy values of jth criteria
d_j	Degrees of divergence values of jth criteria
α_j	Subjective weights of jth criteria
β_j	Objective weights of jth criteria
λ_j	Relational weights coefficients of jth criteria
LB_{ij} or HB_{ij}	Lower the better and higher the batter for ith alternatives on jth criterion
$r(Y_0, Y_i)$	Relative grey relational grades for each of the ith alternatives
W_a	Weights of workpiece after machining in mm^3/min
W_b	Weights of workpiece before machining in mm^3/min
E_a	Weights of tool after machining in g
E_b	Weights of tool before machining in g
t	Machining time in min
ρ	Density of the tool material in g/mm^3
a_{ij}	Average expert opinion values between ith alternatives on jth criterion
A_{ij}	Initial direct relation matrix or average matrix of ith alternatives on jth criterion
I	Identity matrix
K	Total relation matrix
r_j and c_j	Sum of rows and columns for each total relation matrix (K)
T	Weighted decision matrix
I^+ and I^-	Positive and negative ideal solution matrices
V_{ij}	Criteria values of the weighted decision matrix
J	Beneficial criteria

$J^{\text{!`}}$	Non-beneficial criteria
S^+ and S^-	Separation distance of the alternatives
C_i	Closeness values
t_w	Thickness of workpiece
K_t	Top kerf width
K_b	Bottom kerf width
WDM	Weighted decision matrix
W_j	Priority weights for each of the criteria
h	Height of the hole
R	Radius of the hole
R_{tool}	Radius of the tool
G_{ij}	Performance weights of ith alternatives on jth criteria parameters
F	Fuzzy decision matrix
x_j	Criteria weight by the decision maker expressed in triangular fuzzy number
X_{ij}^*	Weighted normalized decision matrix
S_i	Summation of the weighted normalized values of beneficial criteria
M_i	Summation of the weighted normalized values of non-beneficial criteria
M_{min}	Minimum response value in the weighted normalized values of non-beneficial criteria
E_i	Relative significance values of each of the alternatives

Chapter 1
Introduction

1.1 Introduction to Advanced Machining Processes

Manufacturing processes play prominent role in the development of any product. Manufacturing processes include necessary stages for conversion of raw material to finished product that meets the customer specification or expectation. Machining is one type of manufacturing process in which raw materials are cut into desired shape and size through controlled material removal processes. Machining is necessary because different work materials need to be machined with various part shapes and specific geometric features such as screw threads, round holes, straight edges, and surfaces having better surface finish and dimensional accuracy. The process of producing parts through controlled material removal is termed as subtractive manufacturing.

Subtractive manufacturing processes can be classified into conventional or traditional and unconventional or advanced machining processes. Conventional machining processes such as drilling, milling, turning, shaping, grinding, boring, broaching, reaming, honing, lapping, hobbing, polishing are most frequently used for product manufacturing [1]. While advanced machining processes such as electric discharge machining (EDM), wire electric discharge machining (WEDM), electrochemical machining (ECM), ultrasonic machining (USM), laser beam machining (LBM), electron beam machining (EBM), abrasive water jet machining (AWJM), and water jet machining (WJM) have been developed to attain the special machining requirements during product development.

Machining processes where there is a direct contact between the cutting tool and workpiece and where material removal takes through removal of chips, abrasion, or micro-machining are termed as conventional machining [1]. In spite of advantages in machining of various materials, the conventional machining processes possess various disadvantages such as:

- Difficulty in machining of extremely hard and brittle materials,
- Difficulty in machining of intricate shapes and features,

© The Author(s), under exclusive license to Springer Nature Switzerland AG 2019
S. Bhowmik et al., *Modeling and Optimization of Advanced Manufacturing Processes*,
Manufacturing and Surface Engineering, https://doi.org/10.1007/978-3-030-00036-3_1

- Difficulty in machining of parts with high tolerance and surface finish,
- Difficulty in machining of composites, ceramics, and polymers,
- Generation of heat in workpiece and tool.

In order to overcome these disadvantages, advanced processes have been developed and utilized.

Advanced machining processes are special kind of processes where there is no direct contact between the workpiece and tool. In advanced machining processes generally, forms of energies such as electrical energy, kinetic energy, chemical energy are employed for removal of material from the product or workpiece. At present, various advanced machining processes such as EDM, WEDM, USM, ECM, WJM, AWJM, LBM, and EBM are being used frequently in aerospace, automotive, electronics, and other industries due to the significant benefits such as ease in machining of extremely hard materials, complex shapes, parts with less or no heat-affected zone (HAZ), attainment of better surface finish and tolerance compared to the conventional machining process.

1.1.1 Advantages of Advanced Machining Processes

Advanced machining processes are versatile in nature and possess the following as some of the advantages:

- Production of parts with high precision and accuracy.
- Production of intricate and complex shapes.
- Better micro-machining capabilities.
- High production rate while processing difficult-to-machine materials compared to conventional machining processes.
- Machining of material (such as brittle, ductile, hard and non-homogenous) from low to high thickness can be performed by using suitable advanced machining processes.
- Machining of deep holes with a small diameter and with high precision.
- Production of parts with a high surface finish and tight tolerance.

1.1.2 Disadvantages of Advanced Machining Processes

In spite of the versatile features of advanced machining processes, there are a few disadvantages as listed below:

- Comparatively high initial investment.
- Some of the advanced machining processes have very low material removal rate.
- Skilled operators are required for operation of advanced machining processes.
- Electrically non-conductive materials cannot be machined in case of ECM, EBM, and EDM.

1.1.3 Applications of Advanced Machining Processes

Advanced machining processes find applications in almost all fields for machining of a variety of materials with desired product features. The following are some of the major application areas:

- Hard materials for aerospace applications have made the machining by conventional processes difficult and time-consuming, which demands the use of advanced machining processes to overcome those constraints.
- Machining of almost all kind of materials such as metal, rubber, plastic, leather, wood, ceramics, paper, alloys and composites.
- Machining of parts with high precision and accuracy for automobile, telecommunication, computer, and scientific and domestic applications.
- Engraving on glass, metals, parting and machining of precious stones, cutting of parts made of semiconductors and producing fine holes in composites, polymers, and superalloys are possible with advanced machining processes.
- Machining of parts having very low as well as high thickness is possible using advanced machining processes.
- Advanced machining processes are extensively used in medical industry for the production of surgical instruments, bio-implants, and medical devices.

1.2 Classification of Advanced Machining Processes

Depending upon the type of energy used in the process of machining such as mechanical, electrochemical, electrothermal, and chemical, the advanced machining processes can be categorized into different types. The mechanical processes include abrasive jet machining (AJM), water jet machining (WJM), abrasive water jet machining (AWJM), ultrasonic machining, and abrasive flow machining (AFM), while electrochemical processes include electrochemical milling, drilling, and grinding. The electrothermal processes include laser beam machining (LBM), electron beam machining (EBM), electrical discharge machining (EDM), and wire electric discharge machining (WEDM). Whereas, chemical milling and photochemical drilling are chemical processes. The detailed classification of advanced machining processes is shown in Fig. 1.1.

1.3 Modeling and Optimization in Advanced Machining Processes

With the advent of advanced machining processes, it is possible to machine parts having complex shapes, hard materials, composites with better quality, high tolerance, high surface finish and with less or no HAZ. But, in order to obtain these desired

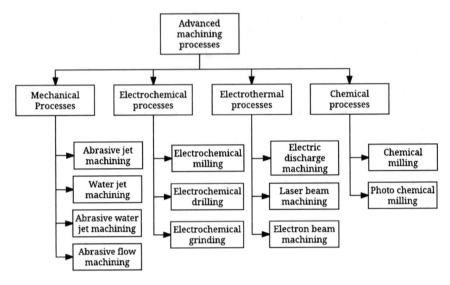

Fig. 1.1 Classification of advanced machining processes

characteristics with low manufacturing cost, better quality, and high efficiency, it is necessary to perform the machining at optimum parametric conditions. It is indeed a difficult task even for a well-experienced operator to know the optimum input parameter conditions to acquire the desired output characteristics. Performing the advanced manufacturing processes at conditions where the parameters are not optimized may lead to more energy consumption, wastage of process time, poor surface finish, less material removal rate, and lack of precision in shape and dimensional accuracy of the product with less productivity. Therefore, it is necessary to make use of appropriate modeling and optimization techniques in advanced machining processes to find the optimum input parametric values to get the best output characteristics.

Since advanced machining processes possess numerous input parameters or factors and the outputs may be multiple, it is necessary to choose an optimization technique that is capable of solving such single- and multi-objective-type complex optimization problems to provide optimum set of machining parameters. Generally, the outputs rely on the significance level of the priority weights of the different parameters. The knowledge and experience of the decision maker play a vital role while assigning priority weights to the response/process parameters. Inappropriate selection of significance level of the response/process parameters may lead to non-optimized results which may have a profound impact on the outcome and efficiency of the advanced machining processes.

Furthermore, ideal selection of process parameters from databook readings is not a recommended approach as it does not give the optimal results. Due to the above reasons, the ideal choice of the process parameters for advanced machining processes is considered to be a complex task for multi-parameter optimization problems. Likewise, it also contains multiple attributes/criteria with uncertain information and subjective + objective data which make the advanced machining process optimization a more intricate and challenging task. Hence, a viable and effective decision-making approach based on multi-criteria decision-making (MCDM) methods is a fundamental prerequisite for powerful modeling and process parameter optimization of advanced machining processes [2].

1.4 Multi-criteria Decision-Making Optimization Techniques

Multi-criteria decision making is a sub-discipline of operations research that is widely used in many fields such as energy planning, construction, environmental projects, software engineering, military applications, food safety, solid waste management, agricultural production, risk management, manufacturing system, and other fields [2]. There is a wide variety of MCDM techniques available and used depending upon the field of application in which it is used. MCDM technique is used to select the best alternative among many alternatives satisfying multiple criteria. In other words, MCDM technique is applied for solving decision and planning problems having multiple criteria which are different in nature [3].

The classifications of various MCDM techniques are given in Fig. 1.2. Some of the most well-known MCDM techniques are analytic hierarchy process (AHP) [4], TOPSIS (Technique for Order Preference by Similarity to Ideal Solution) [5], PROMETHEE, and ELECTRE [5]. But these techniques do not consider the uncertainty issues in the optimization. However, to handle the uncertainty issues, fuzzy logic-based MCDM techniques are employed [6]. The various fuzzy-based MCDM

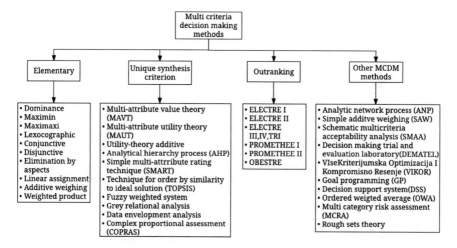

Fig. 1.2 Classification of MCDM techniques [7, 8]

techniques utilized are neuro-fuzzy, fuzzy–GA, and fuzzy–ES, etc. Literature review reveals that the mostly used hybrid techniques in MCDM are fuzzy logic and expert system, fuzzy logic and evolutionary algorithm, and fuzzy logic and neural network.

1.4.1 MCDM Techniques in Advanced Machining Processes

1.4.1.1   Electric Discharge Machining Process

Electric discharge machining (EDM) is a thermal-type advanced machining process where the mechanism of material removal is erosion caused by the electrical discharge occurring between anode and cathode [9]. It occurs between tool electrode (cathode) and electrically conductive workpiece (anode) under the influence of suitable dielectric fluid due to rapid and repetitive spark discharge. The input parameters in EDM are pulse-on time (T_{on}), discharge current, discharge voltage, pulse-off time (T_{off}), etc., and the output responses are material removal rate (MRR), surface roughness (SR), and tool wear rate (TWR), etc.

Since there are numerous input parameters which influence the machining process, proper selection of parameters plays an important role. The percentage use of various optimization techniques for modeling and optimization of EDM process parameters is shown in Fig. 1.3. After analyzing the literature cited in the articles [10–19], it has been found that among 147 research articles on optimization techniques, only 33 articles (i.e., 23% of total number of articles) related to MCDM techniques, while 26 articles (i.e., 79% of the 33 articles) related to individual MCDM techniques such as GRA, fuzzy logic, and DEA, and seven articles (i.e., 21% of the 33 articles) related to hybrid approach such as fuzzy–TOPSIS, grey–fuzzy are utilized for modeling and optimization of EDM-based processes. It can also be observed that among the other MCDM techniques, GRA is used by many researchers with 60.61% (among the 33 articles).

1.4.1.2   Electrochemical Machining Process

Electrochemical machining (ECM) process is the controlled removal of metals by the anodic dissolution [20] in an electrolytic medium, where the workpiece (anode) and the tool (cathode) are connected to the electrolytic circuit, which is kept immersed in the electrolytic medium. The shape of the workpiece obtained is a mirror image of the tool. The input parameters are current (I), feed rate (FR), interelectrode gap, electrolyte flow rate, voltage (V), and the output responses are MRR and surface roughness (SR).

Proper selection of parameters plays an important role as there are a variety of input and output parameters involved in the machining process. An analysis on the implementation of the different optimization techniques for ECM process is shown in Fig. 1.4. Based upon the literature cited and references in the articles [21–26],

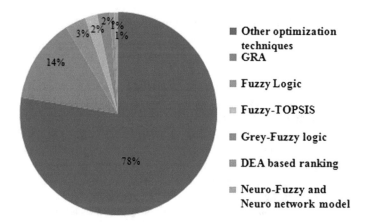

Fig. 1.3 Percentage use of MCDM techniques in EDM-based processes

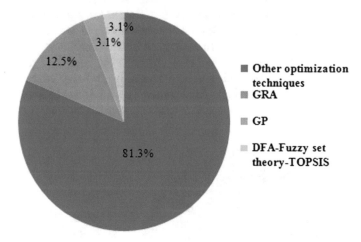

Fig. 1.4 MCDM techniques used in ECM process

among 32 research articles related to various optimization techniques, six articles (i.e., 18.8% of total number of articles) related to MCDM techniques are utilized for the modeling and optimization of ECM process parameters. It has also been observed that the individual techniques (five articles, i.e., 83.33% of six articles) based on MCDM such as GRA and goal programming and hybrid techniques (one article, i.e., 16.67% of six articles) such as DFA–fuzzy set theory–TOPSIS are utilized. It can be observed that GRA is the most used MCDM technique among the researchers with 66.67% (among six articles).

1.4.1.3   Wire Electric Discharge Machining Process

The working principle of wire electric discharge machining (WEDM) process is similar to EDM process. WEDM process is a special type of EDM process in which a small diameter wire is used as the cutting tool. The cutting wire is advanced continuously between the supply spool and wire collector, and the material is removed by a series of electric discharges occurring between the wire electrode and the workpiece. The dielectric fluid serves the dual purpose of flushing away the debris and also acting as a coolant [27]. The area where discharge takes place is heated to a very high temperature and gets melted and removed by the flowing dielectric fluid. The process parameters in WEDM process are wire feed, wire tension, T_{on}, T_{off}, and V, while the output responses are MRR, SR, and kerf width (KW).

Selection of process parameters in WEDM is required to be done with utmost care as there are a number of input and output process parameters. An analysis on the use of different techniques for optimizing the WEDM process is shown in Fig. 1.5. The result shows that among 113 articles (cited and referenced in [28–32]) on modeling and optimization of WEDM process parameters, 11 articles (i.e., 9.7% of the 113 articles) are related to MCDM techniques which comprises of ten individual MCDM techniques (i.e., 90.91% of ten articles) such as GRA and one hybrid MCDM technique (i.e., 9.09% of ten articles), i.e., adaptive neuro-fuzzy inference system with artificial bee colony algorithm. As shown in Fig. 1.5, it can be seen that GRA is the most used MCDM technique with 90.91% (among ten articles).

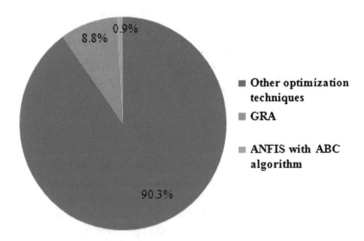

Fig. 1.5 MCDM techniques used in WEDM process

1.4.1.4 Ultrasonic Machining Process

Ultrasonic machining (USM) process is an advanced machining process where material removal from the work surface is done by micro-chipping caused due to repetitive impact of abrasive particles (held in liquid medium) through an ultrasonically vibrated shaped tool [33]. The process parameters are abrasive grain size, tool material, and abrasive concentration, while the general output responses being tool wear ratio (TWR), SR, and MRR.

Figure 1.6 presents the statistics of various techniques used for optimization of USM process. Among 20 articles (cited and discussed in [34–37]) related to optimization of USM process parameters, five articles are related (i.e., 25% of 20 articles) to MCDM techniques where adaptive neuro-fuzzy inference system is the most used MCDM technique.

1.4.1.5 Abrasive Water Jet Machining Process

Abrasive water jet machining (AWJM) process is an extended version of water jet machining in which abrasive particles such as silicon carbide or aluminum oxide are added to water so as to increase the metal removal rate [38]. By using AWJM process, metallic, non-metallic, and composites of various thicknesses can be cut, and it is particularly suitable for heat-sensitive materials which cannot be machined by other processes which produce heat. The AWJM process parameters are pressure, abrasive flow rate, standoff distance, and traverse rate, and the common output responses are kerf geometry, SR, and MRR.

An analysis of the different optimization techniques used in optimizing the AWJM process parameters is shown in Fig. 1.7. Based upon the analysis of the articles cited and referenced in [39–41], it has been observed that among 31 articles related to

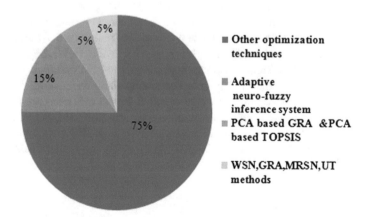

Fig. 1.6 MCDM techniques used in USM process

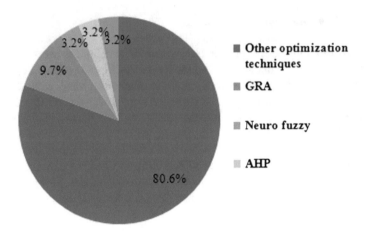

Fig. 1.7 MCDM techniques used in AWJM process

various optimization techniques, six articles (i.e., 19.35% of the 31 articles) are found to be on MCDM techniques [39–41] with four individual MCDM techniques such as GRA and AHP and two hybrid MCDM techniques such as, neuro-fuzzy, ANOVA, and Derringer desirability-based MCDM technique. It is also observed that GRA is the most used MCDM technique with 50% (among the six articles).

1.4.1.6   Laser Beam Machining Process

Laser beam machining (LBM) process is a thermal material removal process that uses a high energy coherent light beam to melt and vaporize the material from the surface of metallic and non-metallic workpiece [42]. Lasers can be used to cut, drill, weld, and mark. LBM process is specifically suitable for producing precise holes with high accuracy. The various parameters that have been considered for optimization, in general, are gas pressure, lamp current, pulse width, pulse frequency, and cutting speed with some common output responses as hole taper, kerf width (KW), SR, and MRR.

An analysis of the different optimization techniques used in optimizing the LBM process parameters is shown in Fig. 1.8 The results show that among 54 articles related to modeling and optimization, 19 research papers (35.19%) are related to MCDM techniques [43–49] with 13 papers on individual MCDM techniques such as GRA and fuzzy logic model, and 15 papers on hybrid MCDM techniques such as grey–Taguchi, GRA coupled with PCA, GRSM, grey–fuzzy, Taguchi-based fuzzy, and RSM–GRA.

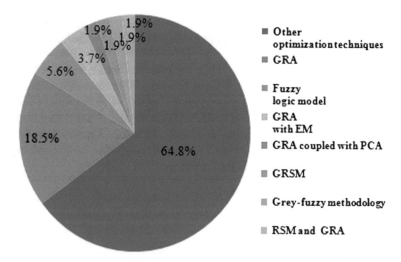

Fig. 1.8 MCDM techniques used in LBM process

1.5 Summary

This chapter provides a brief overview on modeling and optimization of various advanced machining processes based on MCDM techniques and the percentage use of sub-techniques in the advanced processes based upon the literature study. It has been observed that among the various advanced machining processes, modeling and optimization of EDM process parameters using MCDM techniques are attempted by many researchers with 39.3% followed by LBM process with 22.6%, and photochemical machining (PCM) process is the least with 4.8% (Fig. 1.9).

Fig. 1.9 Percentage of MCDM techniques used in advanced machining processes

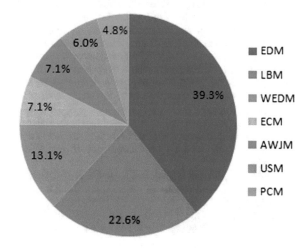

From the analysis of the various kinds of literature, it is also observed that GRA and fuzzy contribute to around 97% among the other MCDM techniques used in all of the advanced machining processes, either as an individual or as a hybrid technique. Furthermore, application of some MCDM techniques such as GP, TOPSIS, DEA, and AHP are also seen for advanced machining processes. The observations encourage further research and development as regards to the use of multi-criteria decision-making techniques for modeling and optimization in advanced machining processes.

The subsequent chapters present the details of the experimental research conducted by the authors on modeling and optimization of advanced machining and manufacturing processes using MCDM techniques.

References

1. Venkata Rao R (2011) Advanced modelling and optimization of manufacturing processes. International Research and Development. Springer, New York
2. Mardani A, Jusoh A, Nor KMD, Khalifah Z, Zakwan N, Valipour A (2015) Multiple criteria decision-making techniques and their applications: a review of the literature from 2000 to 2014. Econ Res-Ekonomska Istraživanja 28(1):16–571
3. Majumder M (2015) Impact of urbanization on water shortage in face of climatic aberrations, Springer Briefs in Water Science and Technology. Springer, New York
4. Saaty TL (1986) Axiomatic foundation of the analytic hierarchy process. Manage Sci 32(7):841–855
5. Hwang CL, Yoon K (1979) Multiple attribute decision making methods and applications. A state-of-the-art survey: Lecture notes in Economics and Mathematical systems. Springer, New York
6. Bellman RE, Zadeh LA (1970) Decision-making in a fuzzy environment. Manage Sci 17(4):B141–B164
7. Zardari NH, Ahmad K, Shirazi SM, Yusop ZB (2015) Weighting methods and their effects on multi-criteria decision making model outcomes in water resources management, Springer Briefs in Water Science and Technology
8. Eshlaghy AT, Homayonfar M (2011) MCDM methodologies and application: a literature review from 1999 to 2009. Res J of Int Stud 21:86–137
9. Abbas NM, Solomon DG, Bahari F (2007) A review of current research trends in electrical discharge machining (EDM). Int J Mach Tools and Manuf 47:1214–1228
10. Lin CL, Lin JL, Ko TC (2002) Optimization of the EDM process based on the orthogonal array with fuzzy logic and grey relational analysis method. Int J Adv Manuf Technol 19(4):271–277
11. Lin JL, Lin CL (2005) The use of grey–fuzzy logic for the optimization of the manufacturing process. J Mater Process Technol 160:9–14
12. Tzeng YF, Chen FC (2007) Multi-objective optimisation of high-speed electrical discharge machining process using a fuzzy-based approach. Mater Des 28:1159–1168
13. Pradhan MK, Biswas CK (2010) Neuro-fuzzy and neural network-based prediction of various responses in electrical discharge machining of AISI D2 steel. Int J Adv Manuf Technol 50:591–610
14. Reddy VC, Sivaiah P, Venkata Chalapathi K, Sangha Mitra RK (2014) Multi-response optimization of EDM process of OHNS using fuzzy logic approach. Int J Res Advent Technol 2(11):9–14
15. Lin JL, Lin CL (2005) The use of grey–fuzzy logic for the optimization of the manufacturing process. J Mater Process Technol 160:9–14

16. Pattnaik SK, Priyadarshini M, Mahapatra KD, Mishra D, Panda S (2015) Multi-objective optimization of EDM process parameters using fuzzy method. In: IEEE sponsored 2nd international conference on innovations in information, embedded and communication systems (ICIIECS)
17. Samantra C, Satyaranjan S, Kumar A (2016) Multi-objective optimization of EDM process parameters using fuzzy based approach. Int J Eng Stud Tech Appr 2(4):2–10
18. Reddy VV, Valli PM, Kumar A, Reddy CHS (2015) Multi-objective optimization of electrical discharge machining of PH17-4 stainless steel with surfactant-mixed and graphite powder—mixed dielectric using—data envelopment analysis—based ranking method. J Eng Manuf 229(3):487–494
19. Lin JL, Lin CL (2002) The use of the orthogonal array with grey relational analysis to optimize the electrical discharge machining process with multiple performance characteristics. Int J Mach Tools Manuf 42:237–244
20. Rajurkar KP, Zhu D, McGeough JA, Kozak J, De Silva A (1999) New developments in electro-chemical machining. CIRP Ann Manuf Technol 48(2):567–579
21. Batra JL, Acharya BG, Jain VK (1986) Multi-objective optimization of the ECM process. Precis Eng 8(2):88–96
22. Santhi M, Ravikumar R, Jeyapaul R (2013) Optimization of process parameters in electro-chemical machining (ECM) using DFA–fuzzy set for titanium alloy. Multidiscipline Model Mater Struct 9(2):243–255
23. Das MK, Kumar K, Barman TK, Sahoo P (2014) Optimization of surface roughness and MRR in electrochemical machining of EN31 tool steel using grey-Taguchi approach. Procedia Mater Sci 6:729–740
24. Chakradhar D, Venu Gopal A (2011) Multi-objective optimization of electrochemical machining of EN31 steel by grey relational analysis. Int J Model Optim 1(2)
25. Manikandan N, Kumanan S, Sathiyanarayanan C (2017) Multiple performance optimizations of electrochemical drilling of Inconel 625 using based grey relational analysis. Eng Sci Technol Int J 20(2):662–671
26. Tang L, Yang S (2013) Experimental investigation on the electrochemical machining of 00Cr12Ni9Mo4Cu2 material and multi-objective parameters optimization. Int J Adv Manuf Technol 67(9–12):2909–2916
27. Gupta K, Jain NK, Laubscher RF (2015) Spark-erosion machining of miniature gears: a critical review. Int J Adv Manuf Technol (Springer) 80(9–12):1863–1877
28. Gupta K, Jain NK (2013) On productivity of wire electric discharge machining for manufacturing of miniature gears. In: Proceedings of 2nd international conference on intelligent robotics, automation and manufacturing 2013 (IRAM 2013), 428–439, Dec 16–18, 2013. Indore, India
29. Jana AK, Akshay R, Anil Reddy R, Vardhan V, DH, Sai Kumar V (2016) Optimization of process parameters in WEDM by using GRA technique for machining Aluminium HE15WP. Int J Eng Innovative Technol (IJEIT) 5(10)
30. Jangra K, Jain A, Grover S (2010) Optimization of multiple-machining characteristics in wire electrical discharge machining of the punching die using grey relational analysis. J Sci Ind Res 69:606–612
31. Huang JT, Liao YS (2003) Optimization of machining parameters of Wire-EDM based on grey relational and statistical analyses. Int J Prod Res 41(8):1707–1720
32. Kumar A, Gulati V, Goswami A (2015) Optimization of process parameter in WEDM for Monel K-500 using Ultima-1F and wire-cut EDM. Int J Res Aeronaut Mech Eng 3(4):53–68
33. Pei ZJ, Ferreira PM, Haselkorn M (1995) Plastic flow in of ceramics. J Mater Process Technol 48(1–4):771–777
34. Gauri SK, Chakravorty R, Chakraborty S (2011) Optimization of correlated multiple responses of (USM) process. Int J Adv Manuf Technol 53:1115–1127
35. Chakravorty R, Gauri SK, Chakraborty S (2013) Optimization of multiple responses of (USM) process: a comparative study. Int J Ind Eng Comput 4:285–296
36. Gill SM, Singh J (2010) An Adaptive neuro-fuzzy Inference system modeling for material removal rate in the stationary ultrasonic drilling of sillimanite. Expert Syst Appl 37(8):5590–5598

37. Singh J, Gill SG (2009) Fuzzy modeling and simulation of ultrasonic drilling of porcelain with hollow stainless steel tools. Mater Manuf Processes 24(4):468–475

38. Phokane TC, Gupta K, Gupta MK (2018) Investigations on surface roughness and tribology of miniature brass gears manufactured by abrasive water jet machining. In: Proceedings of IMechE, Part C: Journal of Mechanical Engineering Science (Sage)

39. Azmir MA, Ahsan AK, Rahmah A, Noor MM, Aziz AA (2007) Optimization of abrasive water jet machining process parameters using orthogonal array with grey relational analysis. In: Regional conference on engineering mathematics, mechanics, manufacturing & architecture 21–30

40. Gaidhani YB, Kalamani VS (2013) Abrasive water jet review and parameter selection by AHP method. IOSR J Mech Civil Eng 8(5):1–6

41. Jegaraj JJR, Babu NR (2007) A soft computing approach for controlling the quality of cut with abrasive water jet cutting system experiencing orifice and focusing tubewear. J Mater Process Technol 185(1–3):217–227

42. Schaeffer R (2012) Fundamentals of laser. CRC Press; Li CH, Tsai MJ (2009) Multi-objective optimization of laser cutting for flash memory modules with special shapes using grey relational analysis. Optics Laser Technol 41:634–642

43. Mishra S, Yadava V (2013) Modeling and optimization of laser beam percussion drilling of thin aluminum sheet. Opt Laser Technol 48:461–474

44. Pandey AK, Dubey AK (2012) based fuzzy logic optimization of multiple quality characteristics in laser cutting of Duralumin sheet. Opt Lasers Eng 50:328–335

45. Pan LK, Wang CC, Wei SL, Sher HF (2017) Optimizing multiple quality characteristics via method-based grey analysis. J Mater Process Technol 182:107–116

46. Syna CZ, Mokhtar M, Feng CJ, Manurung YHP (2011) Approach to prediction of laser cutting quality by employing fuzzy expert system. Expert Syst Appl 38:7558–7568

47. Badkar SD, Pandey KS, Buvanashekaran G (2011) Parameter optimization of laser transformation hardening by using method and utility concept. Int J Adv Manuf Technol 52:1067–1077

48. Sharma A, Yadava V (2011) Optimization of cut quality characteristics during Nd:YAG laser straight cutting of Ni-based superalloy thin sheet using grey relational Analysis with entropy measurement. Mater Manuf Processes 26:1522–1529

49. Allen DM (1986) The principles and practice of photochemical machining and photoetching. Adam Hilger, Techno House, UK

Chapter 2
Modeling and Optimization of Electrical Discharge Machining

2.1 Introduction

Electrical discharge machining (EDM) is one of the important non-traditional machining processes that is employed for machining of DTM materials (such as high strength temperature resistant alloys, composites, and ceramics), micro-machining, and mold–die manufacturing, etc. In EDM process, the workpiece and tool are generally connected to positive and negative terminals, respectively, and immersed in a dielectric fluid (see Fig. 2.1). The tool and workpiece are both conductors of electricity in EDM process. A gap maintained between the tool and workpiece causes potential difference between them. The potential difference between tool and workpiece causes the movement of the electrons and ions, i.e., spark, and that removes material. The EDM process works on electrothermal process principal where kinetic energy of the electrons and ions is converted into thermal energy and causes high temperature due to bombardment of spark that results in removal of material due to the instant melting and vaporization [1–3]. The adverse effects of thermal energy may lead to high tool wear, surface quality deterioration, and slow material removal rate.

Voltage, current, pulse-on time, pulse-off time, electrode material and diameter, and lifting height of the electrode are the important input parameters of the EDM process. While material removal rate directly relates to productivity, surface roughness parameters such as average roughness (which is arithmetic average of the absolute values of the roughness profile ordinates) give clear indication of the surface quality of the machined sample; electrode and tool wear rate are the important output parameters in EDM process.

In addition, the EDM process releases a great amount of toxic substances in the form of solid, liquid, and gaseous wastes, resulting in serious occupational health and environmental issues. Apart from the generation of waste, the manufacturing process is considered to be an energy-intensive activity which may indirectly affect the environment. Therefore, the amount of waste generated and its performance parameters

© The Author(s), under exclusive license to Springer Nature Switzerland AG 2019
S. Bhowmik et al., *Modeling and Optimization of Advanced Manufacturing Processes*,
Manufacturing and Surface Engineering, https://doi.org/10.1007/978-3-030-00036-3_2

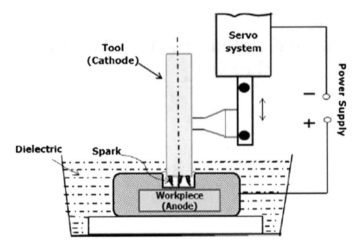

Fig. 2.1 Working principle of electric discharge machining process

are strongly influenced by the process parameters. Hence, modeling and optimization of the EDM process is essential in order to achieve optimal values of responses such as higher material removal rate, good surface roughness, lower electrode wear rate. The modeling and optimization of EDM process is further considered to be a multi-criteria decision-making (MCDM) problem.

Many investigations have been attempted for modeling and optimization of the EDM process by various MCDM techniques such as analytical hierarchical process [4–6], grey relational analysis [6, 7], VIKOR [6, 8], technique for order preference by simulation of ideal solution (TOPSIS) [9, 10], compromised weighted method, multi-objective optimization on the basis of ratio analysis (MOORA) [11–13], and multi-objective optimization on the basis of simple ratio analysis (MOOSRA) [12]. However, these methods do not consider the priority weights (both subjective, i.e., uncertain, imprecise, and vagueness information, and objective weights) information, and the percentage of contribution of each input parameters and interrelationship aspects of the parameters during the modeling which lead to error in the optimal results. Considering the drawbacks of these methods, in the present chapter, an integrated MCDM method for modeling and optimizing the EDM process parameter has been attempted. The Integrated MCDM techniques include subjective + objective weighted (SOW) methods with weighted grey relational analysis (WGRA). Here, the priority weights of the parameters are considered by subjective + objective weighted (SOW) method while percentage contribution and interrelationship aspects of the parameters are considered by weighted grey relational analysis (WGRA) [8].

2.2 Integrated SOW–WGRA-Based MCDM Method

The integrated MCDM method consists of both subjective + objective weighted (SOW) method [14, 15] and weighted grey relational analysis (WGRA) [8]. The detailed steps of the integrated MCDM method implemented are explained as follows:

Step 1: Development of the Decision Matrix

In modeling and optimization, the development of decision matrix consists of performance values of the criteria as output parameters and corresponding alternatives as process/input parameters settings. The decision matrix is formulated using the following expression as given in Eq. (2.1).

$$D_{ij} = \begin{bmatrix} & C_1 & C_2 & \cdots & C_n \\ A_1 & Y_{11} & Y_{12} & \ldots & Y_{1j} \\ A_2 & Y_{21} & Y_{22} & \cdots & Y_{2j} \\ \vdots & \vdots & \vdots & \cdots \cdots & \vdots \\ A_i & Y_{i1} & Y_{i2} & \cdots & Y_{ij} \end{bmatrix} \quad \text{for } i = 1, 2, 3, 4 \ldots m \quad \text{and} \quad j = 1, 2, 3, 4 \ldots n$$

$$(2.1)$$

where Y_{ij} is the performance measure of ith alternatives/experimental runs or input parameters setting on jth criteria/output parameters, m is the number of alternatives/input parameter settings and n is the number of criteria/output parameters, A_1, A_2, A_3, ... A_i represent alternatives/input parameter, C_1, C_2, C_3 ... C_j represent criteria/output parameters, and D_{ij} is the decision matrix.

Step 2: Determination of Precise Weights of Each Criteria/Output Parameters

In this step, precise priority weights of each criteria/response parameters are determined using subjective + objective weighted (SOW) method. First, subjective weights for each of the criteria/output parameters are calculated via fuzzy theory. Fuzzy decision matrix is developed taking opinion of the decision makers. Each of their opinion for each of the criteria/output parameters is tabulated in terms of linguistic variables using triangular fuzzy numbers (TFNs). Then, assigned weights for each of the criteria/output parameters are converted into fuzzy subjective weights (α_j) via aggregation of triangular fuzzy numbers.

Second, the objective weights are assigned for each of the criteria/output parameters by entropy method. Here, first, decision matrix is formulated which involves number of criteria/output parameters and their corresponding performance values. Second, normalization of the decision matrix is carried out in order to convert the different units of the performance values into comparable sequence using Eq. (2.2).

$$E_{ij} = \frac{D_{ij}}{\sqrt{\sum_{i=1}^{m} D_{ij}^2}} \quad (i = 1, 2, \ldots, m; \ j = 1, 2, \ldots, n) \qquad (2.2)$$

where D_{ij} is the response values of the ith alternative to the jth criteria/output parameters, m denotes the number of alternatives/process parameter settings and n represents the number of criteria/output parameters. Then, the entropy value (p_j), degrees of divergence (d_j), and entropy/objective weights (β_j) for each of the criteria/output parameters are calculated using the following expressions.

$$p_j = -k \sum_{i=1}^{m} E_{ij} \ln(E_{ij}) \quad j = 1, 2 \ldots n \qquad (2.3)$$

$$d_j = \left| 1 - p_j \right| \qquad (2.4)$$

$$\beta_j = \frac{d_j}{\sum_{j=1}^{n} d_j} \qquad (2.5)$$

where $k = 1/\ln(m)$ is a constant term, its value in the range between $0 \leq p_j \leq 1$, and mm denotes the number of alternatives/input parameter settings, d_j denotes the degrees of divergence values of each criteria/output parameters, and β_j is the objective weights or entropy weights of the jth criteria/output parameters.

After the evaluation of both subjective + objective weights, rational weights coefficients for jth criteria/output parameters are determined using Eq. (2.6).

$$\lambda_j = \frac{\alpha_j \times \beta_j}{\sum_{j=1}^{n} \alpha_j \times \beta_j} \quad j = 1, 2 \ldots n \qquad (2.6)$$

where p_j is the weight of jth criteria/output parameters obtained via triangular fuzzy method [16–18] and β_j is the weight of jth criteria/output parameters obtained through entropy method [13, 14], α_j is the subjective weights for each of the jth criteria/output parameters, β_j is the objective weights for each of the jth criteria/output parameters, and λ_j is the relational weights coefficients for each of the jth criteria/output parameters.

Step 3: Formulation of Weighted Normalization Matrix

Weighted normalization matrix is formulated by multiplying each performance values of the criteria/output parameters with corresponding precise priority weights obtained via subjective + objective weighted (SOW) method [8]. Generally, three types of normalization process are carried out to render the performance values of the criteria/output parameters, whether the lower is the better (LB), the higher is the better (HB), or nominal the best (NB). The normalized performance matrix of n criteria/output parameters and m alternatives/input parameter settings is obtained using the following equations.

For higher the better

$$HB_{ij} = \lambda_j \times \left(\frac{Y_{ij} - \text{Min}(Y_{ij})}{[\text{Max}(Y_{ij}) - \text{Min}(Y_{ij})]} \right) \quad \text{for} \, (i = 1, 2, \ldots m, \, j = 1, 2, \ldots n)$$

(2.7)

For lower the better

$$LB_{ij} = \lambda_j \times \left(\frac{\text{Max}(Y_{ij}) - Y_{ij}}{[\text{Max}(Y_{ij}) - \text{Min}(Y_{ij})]} \right) \quad (i = 1, 2, \ldots m, \, j = 1, 2, \ldots n)$$

(2.8)

where Y_{ij} is the response value of the jth criteria/output parameters of ith alternatives/input parameter settings; if the normalized (LB_{ij} or HB_{ij}) value of jth criteria/output parameters of ith alternatives/input parameters is equal to 1 or nearer to 1, then it is said that the performance of that particular ith alternatives/input parameters settings is the best for the jth criteria/output parameters and that normalized (Y_{0j}) value is termed as reference value for jth criteria/output parameters. $\text{Min}(Y_{ij})$ is the minimum normalized performance value and $\text{Max}(Y_{ij})$ is the maximum normalized performance value.

Step 4: Evaluation of Equivalent Weighted Grey Relational Coefficients

After the weighted normalization matrix, the weighted grey relational coefficients for each of the criteria/output parameters are evaluated to determine the closeness of normalized (LB_{ij} or HB_{ij}) value or preference normalized (Y_{0j}) value. Higher value of weighted grey relational coefficient means normalized (LB_{ij} or HB_{ij}) value is close to reference normalized (Y_{0j}) value. The weighted grey relational coefficients for each of the criteria/output parameters are calculated using the following expression.

$$\rho(Y_{0j}, \, HB_{ij}/LB_{ij}) = \frac{(\Delta_{\text{Min}} + \xi \Delta_{\text{Max}})}{(\Delta_{ij} + \xi \Delta_{\text{Max}})} \quad (i = 1, 2, \ldots m, \, j = 1, 2, \ldots n) \quad (2.9)$$

where $\rho(X_{0j}, \, HB_{ij}/LB_{ij})$ is the weighted grey relational coefficients for each of the criteria/output parameters, $\Delta_{ij} = |Y_{0j} - Y_{ij}|$ is the degree of closeness values of each criteria/output parameters from the reference values, $\Delta_{\text{Min}} = \text{Min}\{\Delta_{ij}, i = 1, 2, 3, \ldots m; j = 1, 2, 3, \ldots n\}$ is minimum degree of closeness values of each criteria/output parameters and $\Delta_{\text{Max}} = \text{Max}\{\Delta_{ij}, i = 1, 2, 3, \ldots m; j = 1, 2, 3, \ldots n\}$ maximum degree of closeness values of each criteria/output parameters, ξ is the identification coefficient, and the value of identification coefficient (ξ) is taken in the range between $0 \leq \xi \leq 1$.

Step 5: Evaluate the Relative Grey Relational Grades

After the calculation of weighted grey relational coefficients, the overall evaluation of criteria/output parameters to assess the optimal selection of alternatives/process parameter settings is carried out via relative grey relational grades. The relative grey relational grades values for ith alternatives/input parameter settings convert the multi-response parameter problem or multi-criteria problem into single response or criteria problem subjected to the constraints using Eq. (2.10). The constraints are consideration of both beneficial and non-beneficial parameters and interdependencies between the criteria/output parameters and alternatives/input parameter settings.

$$\Gamma(Y_0, \ Y_i) = \sum_{j=1}^{n} \rho(Y_{0j}, \ Y_{0j}, \ \mathrm{HB}_{ij}/\mathrm{LB}_{ij}) \quad \text{for } i = 1, \ 2, \ 3 \ldots m. \qquad (2.10)$$

where $\mathrm{r}(Y_0, \ Y_i)$ is relative grey relational grades for each of the alternatives, and n is the number of alternatives/experimental runs/input parameter setting.

In the final stage, ranking of the alternative/input parameter setting is done based on the relative grey relational grades. The alternative with highest relative grey relational grades yields the optimal rank among other alternatives/input parameter settings.

2.3 Experimental Study

2.3.1 Background

In this section, a case of experimental research conducted by the authors is presented to demonstrate the modeling and process parameter optimization of EDM process using integrated MCDM method, i.e., SOW–WGRA method. The work material used in EDM process is Ti-6Al-4 V alloy. Furthermore, the EDM process performance is based on its input parameters such as discharge current, gap voltage, ratio of pulse width to pulse interval, lifting height. The error in selection of process/input parameters results in error in the optimal values such as lesser material removal rate (MRR), higher electrode tool wear (EWR), poor surface roughness (SR) of the output parameters which directly or indirectly affect the EDM process performance as well as efficiency of the EDM process [1–3].

Further, process parameter optimization of EDM process is considered to be a multi-attribute or multi-criteria decision-making (MCDM) optimization problem due to a wide range of input and output parameters. Therefore, in the present chapter, an attempt is made to implement integrated MCDM method, i.e., SOW–WGRA method, for modeling and optimization of EDM process. Here, subjective + objective weighted (SOW) method is used for extraction of precise priority weights of the criteria/output parameters while selection of optimal alternatives or input/process parameter setting using weighted grey relational analysis (WGRA) method. The outcome of this experimental analysis and optimization is the optimal combination of process parameters for practical applications.

(a)   **(b)**

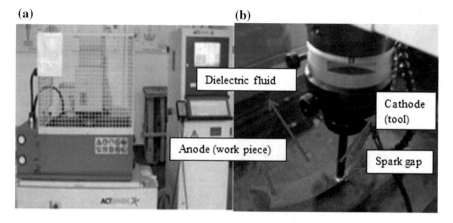

Fig. 2.2 a EDM experimental setup. **b** Machining close view

Table 2.1 Input parameters and their levels

Input parameters	Symbol	Unit	Level 1	Level 2	Level 3
Discharge current	DC	A	11	16	20
Pulse width (T_{on})/pulse interval (T_{off})	T_{on}/T_{off}	μs	30:70	50:50	70:30
Gap voltage	GV	V	20	25	30
Lifting height	LH	mm	3	6	9

2.3.2 Experimental Details

Rectangular workpieces of Ti-6Al-4 V alloy with the dimension of $50 \times 50 \times 8$ mm are used as work specimen for the experimentation. The Ti-6Al-4 V alloy consists of Ti-89.46%, Al-6.08%, V-4.02%, Fe-.22%, O-0.08%, C-0.02%, N-0.01%, and H-0.0053%. The experiments have been conducted using CNC ACTSPARK EDM (Fig. 2.2) with copper electrode of diameter 10 and 100 mm length. Tap water is considered as dielectric medium throughout the experimentation. The input parameters, namely discharge current (DC), pulse width/pulse interval (T_{on}/T_{off}), gap voltage (GV), and lifting height (LH), with three levels each are considered as shown in Table 2.1. The criteria/response parameters, namely material removal rate, electrode wear ratio, and average surface roughness, are considered as shown in Table 2.2.

Based on Taguchi (L9) orthogonal array, nine experiments have been carried out and corresponding response parameters are determined. The response parameters like material removal rate values and electrode wear ratio are calculated using the following expressions.

$$\text{MRR} = \frac{W_b - W_a}{t} \qquad (2.11)$$

Table 2.2 Experimental parameters and results

Exp. no.	Input parameters				Output parameters		
	DC	T_{on}/T_{off}	GV	LH	Material removal rate (MRR) (mm^3/min)	Electrode wear rate (EWR) (mm^3/min)	Surface roughness (SR) (μm)
1	11	30:70	20	3	2.96	0.21	2.17
2	11	50:50	25	6	1.28	0.14	2.37
3	11	70:30	30	9	1.73	0.16	2.83
4	16	30:70	25	9	3.27	0.3	2.19
5	16	50:50	30	3	4.3	0.3	2.61
6	16	70:30	20	6	4.07	0.28	2.86
7	20	30:70	30	6	5.9	0.41	2.15
8	20	50:50	20	9	6.62	0.41	2.65
9	20	70:30	25	3	6.36	0.41	4.27

$$\text{EWR} = \frac{E_b - E_a}{t \times \rho} \tag{2.12}$$

where W_b is weight of the workpiece before machining in mm^3/min, W_a is weight of the workpiece after machining in mm^3/min, t is the machining time in minutes, E_b is the weight of the tool before machining in gram (g), E_a is the weight of the tool after machining in gram (g), and ρ is density of the tool material in g/mm^3. At last, average surface roughness values are measured using surface profile meter. The experimental design and results are given in Table 2.2.

2.4 Modeling of EDM Process

In modeling of EDM, first, decision matrix is formulated using the Taguchi (L9) orthogonal array using Eq. (2.1). The decision matrix includes number of criteria as response parameters and corresponding alternatives as number of experimental runs or process parameter setting. The parameters such as material removal rate (MRR), electrode wear ratio (EWR), and surface roughness (SR); and discharge current (DC), pulse width/pulse interval (T_{on}/T_{off}), gap voltage (GV), and lifting height (LH) are considered as criteria and alternatives, respectively. The criteria, material removal rate (MRR) is considered as beneficial criteria/output parameters, i.e., higher the better, while electrode wear ratio (EWR) and surface roughness (SR) are considered as non-beneficial criteria/output parameters, i.e., lower the better.

Second, determination of the precise priority weights for each of the criteria/output parameters of the EDM process has been carried out using subjective + objective weighted (SOW) method. In this method, initially, subjective weights for each of

Table 2.3 Linguistic variables

Importance	Fuzzy weight
Extremely low (EL)	(0, 0, 0.1)
Very low (VL)	(0, 0.1, 0.3)
Low (L)	(0.1, 0.3, 0.5)
Medium (M)	(0.3, 0.5, 0.7)
High (H)	(0.5, 0.7, 0.9)
Very high (VH)	(0.7, 0.9, 1)
Extremely high (EH)	(0.9, 1, 1)

Table 2.4 Fuzzy weights of criteria/response parameters

Output parameter	Decision maker				Triangular fuzzy weights	Aggregated fuzzy subjective (α_j) weights
	DM1	DM2	DM3	DM4		
MRR	H	H	VH	H	0.55, 0.75, 0.92	0.7417
EWR	VL	L	L	VL	0.05, 0.20, 0.40	0.2167
SR	M	H	M	L	0.75, 0.925, 1.00	0.8917

criteria/output parameters are calculated via fuzzy theory. For this, fuzzy decision matrix is developed taking opinion of the decision makers and four decision makers are used for development of fuzzy decision matrix. Each of their opinion for each of the criteria or output parameters is tabulated (Table 2.3) in terms of linguistic variables using triangular fuzzy numbers.

Then, assigned weights for each of the criteria/output parameters are converted into fuzzy subjective weights (α_j) via aggregation of triangular fuzzy numbers. The results of subjective weights for each of the criteria/output parameters are listed in Table 2.4. The results show that criteria/output parameters, i.e., surface roughness (SR), is found to be more significant compared to the criteria/output parameters material removal rate (MRR) and electrode wear ratio (EWR).

Thereafter, entropy method has been used to compute the objectives weights (β_j) of the criteria/output parameters using Eqs. (2.2)–(2.5). Here, priority weights for each of the criteria/output parameters are calculated based on the decision matrix, and the decision matrix is taken directly from the experimental results as shown in Table 2.2. Then, performance values of the criteria/output parameters are normalized using Eq. (2.2), in order to convert the original sequence data of the criteria/output parameters into comparable sequence data of the criteria/output parameters. Simultaneously, entropy (p_j), degree of divergence (d_j), and objective weights are determined using Eqs. (2.3)–(2.5). The results of objective weights for each of the criteria/output parameters are shown in Table 2.5. Finally, rational weights coefficients (γ_j) for each of the criteria/output parameters are determined using Eq. (2.6) and results are shown in Table 2.5. The results show that the criteria/output parameters, i.e., material removal rate (MRR) as 0.4821, yields the highest relative weight coefficient followed by surface roughness (SR) as 0.4799 and electrode wear ratio (EWR) as 0.0379 for the

Table 2.5 Criteria weights

Methods	MRR	EWR	SR
Subjective weights	0.7410	0.2166	0.8917
Objective weights	0.4770	0.1283	0.3946
Rational weights coefficients	0.4821	0.0379	0.4799

Table 2.6 Grey relational coefficients (GRCs)

Exp. no.	Material removal rate (MRR)	Electrode wear rate (EWR)	Surface roughness (SR)
1	0.552	0.9114	0.4300
2	0.4285	0.9051	0.4569
3	0.4501	0.9114	0.6931
4	0.5440	0.9588	0.4340
5	0.6328	0.9588	0.4929
6	0.6096	0.9518	0.5306
7	0.8475	1	0.4296
8	1	0.9451	0.4846
9	0.9379	0.8540	1

EDM process. This shows that the criteria/output parameters, i.e., material removal rate (MRR) is found to be more significant effect than surface roughness (SR) and electrode wear ratio (EWR) on the performance EDM process.

Similarly, the weighted normalized decision matrix and weighted grey relational coefficients (WGRCs) for each of the criteria/output parameters are determined using Eqs. (2.7)–(2.9), and the results are shown in Table 2.6. The WGRC for each of the criteria/output parameters is evaluated in order to find out the closeness of each criteria/output parameters with respect to the reference point or values.

Thereafter, the overall evaluation values via grey relational grades for each of the criteria/response parameters are determined using Eq. (2.10). The overall evaluation values convert the multi-response parameter or multi-criteria decision-making problem into single response or single criteria decision-making optimization problems subjected to the constraints. The constraints are consideration of both beneficial and non-beneficial parameters and interdependencies between the criteria/process parameters and alternatives/response parameter. In the final stage, ranking of the alternatives/process parameter setting is done based on the relative grey relational grades. The alternative with highest relative grey relational grades yields the optimal rank among other alternatives or experimental setting for the EDM process and the results are shown in Table 2.7.

Table 2.7 Ranking of the alternatives based on grey relational grades

Exp. no.	Input parameters				Output parameters	Rank
	DC	T_{on}/T_{off}	GV	LH	Grey relational grade	
1	11	30:70	20	3	0.2640	2
2	**11**	**50:50**	**25**	**6**	**0.3058**	**1**
3	11	70:30	30	9	0.2460	4
4	16	30:70	25	9	0.2520	3
5	16	50:50	30	3	0.1928	6
6	16	70:30	20	6	0.1806	7
7	20	30:70	30	6	0.2231	5
8	20	50:50	20	9	0.1665	8
9	20	70:30	25	3	0.1129	9

2.5 Results and Discussion

2.5.1 Optimum Combination of Process Parameters

The optimal alternatives or process parameters for the EDM process are selected based on the values of grey relational grades obtained via hybrid MCDM method. The optimized results of EDM process are shown in Table 2.7. The results show that experiment number 2 and their corresponding optimal setting, i.e., process parameters, are optimal among the other experimental runs or process parameter settings. The optimal settings obtained are discharge current (DC) as 11 A, pulse width/pulse interval (T_{on}/T_{off}) as 50:50 μs, gap voltage (GV) as 25 V, and lifting height (LH) as 6 mm. The optimal setting, i.e., experiment number 2, gives optimum criteria/response parameters' values, namely material removal rate (MRR) as 1.28 mm^3/min, electrode wear ratio (EWR) as 0.14 mm^3/min, and surface roughness (SR) as 2.37 μm for the EDM process.

In addition, optimal setting helped to reduce the process time and minimize waste, pollution, and environmental impact by using tap water as dielectric medium; increase the material removal rate (MRR); improve the quality; and improve the efficiency of the EDM process during machining of Ti-6Al-4 V alloy under tap water. Thus, it concludes that the integrated MCDM method, i.e., subjective + objective weighted (SOW) method with weighted grey relational analysis (WGRA), can be used as systematic framework model for modeling and optimization of EDM process on machining of Ti-6Al-4 V alloy.

Table 2.8 Average grey relational grades

Input parameters	Average grey relational grades			Max − Min	Rank
	Level 1	Level 2	Level 3		
DC	0.2719[a]	0.2085	0.1675	0.1044	1
T_{on}/T_{off}	0.2464[a]	0.2217	0.1798	0.0665	2
GV	0.2037	0.2236[a]	0.2206	0.0199	4
LH	0.1899	0.2365[a]	0.2215	0.0466	3

[a]Optimal level

2.5.2 Optimum Input Parameter and Its Level Combination

Additionally, this chapter also analyzes the response of input parameters on the grey relational grades (GRGs) by determining the average GRG for each level of input parameters and the results are shown in Table 2.8. Since the GRG represents the level of correlation between the reference and comparability sequence, comparability sequence has higher GRG for optimum parameters which are discharge current (DC) at level 1, pulse width/pulse interval (T_{on}/T_{off}) at level 1, gap voltage (GV) at level 2, and lifting height (LH) at level 2.

Furthermore, most influencing input parameters are evaluated via difference between maximum and minimum values of the mean GRG and comparing these values. The values obtained are 0.1044 for discharge current (DC), 0.0665 for pulse width/pulse interval (T_{on}/T_{off}),0.0466 for gap voltage (GV), and 0.0466 for lifting height (LH). It is observed from the result that parameter discharge current (DC) found to be more influencing input parameters on the multiple responses of EDM process. The order of importance of the input parameters to the multi-response in the EDM process can be listed as discharge current (DC)—pulse width/pulse interval (T_{on}/T_{off})—lifting height (LH)—gap voltage (GV).

Additionally, the influence of each input parameters can also clearly be represented by response graph as shown in Fig. 2.3. This shows that small changes in the GRG values when a factor goes from their level 1 to 3 and greater GRG values provide the optimum values for the EDM process.

2.6 Summary

In the present chapter, modeling and optimization of the process parameters in EDM on machining of Ti-6Al-4 V alloy is described using integrated MCDM method. The integrated MCDM method comprises subjective + objective weighted (SOW) coupled with weighted grey relational analysis (WGRA) methods. The study concludes

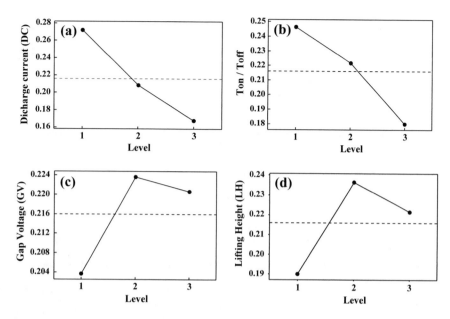

Fig. 2.3 a–d Response graphs of GRG

that use of integrated MCDM method yields the optimal solution and machining of Ti-6Al-4 V alloy is feasible in EDM process. The experiment number 2 is found as the optimal experimental runs among the other runs for EDM process.

The optimal settings obtained are discharge current (DC) as 11 A, pulse width/pulse interval (T_{on}/T_{off}) as 50:50 μs, gap voltage (GV) as 25 V, and lifting height (LH) as 6 mm. The optimal setting, i.e., experiment number 2 gives optimum criteria/response parameters' values, namely MRR(1.28 mm^3/min), EWR (0.14 mm^3/min), and SR (2.37 μm) for the EDM process. In addition, optimal setting improves the performance and efficiency of the EDM process during the machining of Ti-6Al-4 V alloy under tap water. Thus, the present chapter concludes that the integrated MCDM method, i.e., SOW–WGRA method, can be used as efficient standard model for modeling and optimization of various advanced machining processes.

References

1. McGeough JA (1988) Advanced methods of machining. Chapman and Hall, London
2. Davim PJ (2013) Nontraditional machining processes. Springer, London, UK
3. Gupta K, Jain NK, Laubscher RF (2016) Hybrid machining processes. Springer, Berlin
4. Saaty TL (1989) Decision making with the analytic hierarchy process. Int J Serv Sci 1(1):83–98
5. Saket S, Purbey V, Jagadish RA (2014) Multi attributes decision making for mobile phone selection. Int J Res Eng Technol 03(03):497–501

6. Hwang CL, Yoon K (1981) Multiple attribute decision making, a state of the art survey. Springer, New York
7. Tong LI, Chen C, Wang CH (2007) Optimization of multi-response processes using the VIKOR method. Int J Adv Manuf Technol 31:1049–1057
8. Deng JL (1989) Introduction to grey system. J Grey Syst 1(1):1–24
9. Kumar R, Jagadish RA (2013) Selection of material: a multi-objective decision making approach. In: Proceeding of ICIE-2013 international conference on industrial engineering, pp 162–165
10. Kumar R, Jagadish RA (2014) Selection of cutting tool materials: a holistic approach. In: Presented at the 1st international conference on mechanical engineering emerging trends for sustainability, pp 447–452
11. Kumar Rajnish, Jagadish AR (2014) Selection of material for optimal design using multi-criteria decision making. Procedia Mate Sci 6:590–596
12. Jagadish RA (2014) Green cutting fluid selection using MOOSRA method. Int J Res Eng Technol 03(03):559–563
13. Caliskan H, Kursuncu B, Kurbanoglu C, Guven SY (2013) Material selection for the tool holder working under hard milling conditions using different multi criteria decision making methods. Mater Design 45:473–479
14. Emma M, Kieran S, Vida M (2013) An assessment of sustainable housing affordability using a multiple criteria decision making method. Omega 41:270–279
15. Tan XC, Liu F, Cao HJ, Zhang H (2002) A decision-making framework model of cutting fluid selection for green manufacturing and a case study. J Mater Process Technol 129:467–470
16. Tang L, Du YT (2014) Experimental study on green EDM in tap water of Ti-6Al-4 V and parameters optimization. Int J Adv Manuf Technol 70:469–475
17. Zadeh LA (1965) Fuzzy sets. Inf Control 8(3):338–353
18. Bortolan G, Degami R (1985) A review of some methods for ranking fuzzy subset. Fuzzy Sets Syst 15(1):1–19

Chapter 3
Modeling and Optimization of Abrasive Water Jet Machining Process

3.1 Introduction

Abrasive water jet machining (AWJM) process is a type of non-traditional machining process in which the mechanical energy of water and abrasive particles are used to achieve the material removal or machining [1]. In abrasive water jet machining (AWJM) process, the material removal takes place through controlled erosion via high-velocity abrasive particles impinging on the workpiece (see Fig. 3.1). An abrasive water jet machining system consists of four major components such as pumping system, abrasive feed system, water jet and abrasive jet nozzle [1].

The various input parameters such as water jet pressure, abrasive grain size, abrasive mass flow rate, nozzle traverse speed, standoff distance affect the performance of abrasive water jet machining process.

Material removal rate (MRR) which is directly related to the process productivity, surface roughness which depicts the surface quality and determines the functional performance of the machined part, and geometric accuracy (part waviness, form features, kerf, and taper, etc.) are the major output parameters in AWJM process [1].

Improper selection of any of the parameter leads to increase in the cost of manufacturing, cycle time as well as a decrease in quality of the product with less productivity. In general, selection of process parameters is done based on the experience and expertise of the operator or from the handbooks. However, selected parameters are not optimal because knowledge of operator and sometimes the values of the parameters are obtained to be far from optimum conditions.

In spite of the above reasons, it is costly, time-consuming, and tedious to determine the ideal values of process parameters without optimization. Furthermore, nature of the output responses of abrasive water jet machining process is contradictory, i.e., metal removal rate, process energy being higher the better, while surface roughness,

Fig. 3.1 Working principle
of AWJM process

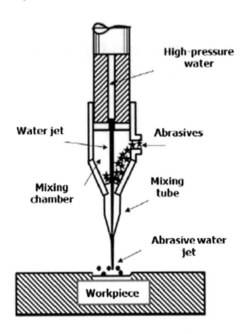

taper angle, and process time being lower the better. This leads to the selection of optimal process parameters for abrasive water jet machining process more complex and challenging task and also considered to be multi-response parameter optimization or multi-criteria decision-making (MCDM) optimization problem. Thus, multi-criteria decision-making (MCDM) method for modeling and optimum selection of these process parameters is essential.

In fact, many investigations have been conducted on optimization of process parameters of AWJM process using MCDM techniques, such as grey relational analysis with Taguchi method [2, 3], analytical hierarchical process (AHP) [4], neuro-fuzzy [5], ANOVA with Derringer–Suich multi-criteria optimization technique [6], and VIKOR [7]. It has been observed that very few MCDM methods were employed for modeling and optimization of AWJM process. Likewise, considerations of precise priority weights as well as subjective + objective data were omitted by existing methods during AWJM process optimization. In addition, the effects of favorable and detrimental parameters for each of the response parameters were not taken into account during the ranking of process parameters. To overcome the above issues, suitable decision-making model based on multi-criteria decision-making techniques is essential. Hence, the present chapter has put forward an integrated MCDM method, i.e., modified-decision-making and trial evaluation laboratory (DEMATEL)–technique for order preference by similarity to ideal solution (TOPSIS) for modeling and optimization of AWJM process parameters.

3.2   Integrated M-DEMATEL–TOPSIS-Based MCDM Method

The integrated MCDM method which consists of M-DEMATEL [8, 9] combined with TOPSIS [10] is discussed. The comprehensive steps of DEMATEL–TOPSIS methodology are as follows.

Step 1: Establishing Decision Matrix

The modeling and optimization of AWJM process begins with the establishment of decision matrix (D_{ij}) which consists of alternatives/input parameters and their corresponding criteria values/output parameters. The number of rows in the decision matrix is equal to the number of experimental runs or input parameter settings obtained using Taguchi orthogonal array, and the number of columns is equal to the number of criteria/output parameters considered. The criteria can be a combination of qualitative and quantitative criterion which has significant influence on the performance of the process. The entries of D_{ij} matrix are each criteria/parameters' values corresponding to different process parameter setting as per the Taguchi array. Formulation of decision matrix is done with the following equation.

$$D_{ij} = \begin{bmatrix} & C_1 & C_2 & \dots & C_n \\ A_1 & Y_{11} & Y_{12} & \dots & Y_{1n} \\ A_2 & Y_{21} & Y_{22} & \dots & Y_{2n} \\ \dots & \dots & \dots & \dots & \dots \\ A_m & Y_{m1} & Y_{m2} & \dots & Y_{mn} \end{bmatrix} \tag{3.1}$$

where C_1, C_2, … C_n represent the number of criteria/output parameters, A_1, A_2, … A_n denote the number of alternatives/input parameters, and Y_{11}, Y_{12} … Y_{mn} denote the performance value of each criterion/output parameter corresponding to the alternatives/input parameter setting; m is the number of the criteria/output parameters, and n is the number of alternatives/experimental runs or input parameters.

Step 2: Calculation of Normalized Decision Matrix

Normalization of the decision matrix is carried out in order to convert the criteria/output parameters which are in different units into dimensionless terms. Since comparison and assigning between two different criteria are not feasible as long as they are in different units, normalization is performed. The normalized values denote the relative performance of the generated design criteria/output parameters. The normalization of the different criteria/output parameters is done using the following equation.

$$N_{ij} = \frac{Y_{ij}}{\sqrt{\sum_{i=1}^{m} Y_{ij}^2}} \tag{3.2}$$

where N_{ij} represents the normalized values of each of the criteria (j)/output parameters with respect to the alternatives (i)/input parameters.

Step 3: Determination of Average Matrix

Then, the average matrix or initial direct relation matrix is formulated using Eq. (3.3).
In this, first, a number of experts (H) are considered for the formulation of the comparative decision matrix. Second, each expert (H) assigns the degree of importance
between two criteria $(i$ and $j)$ using the integer scale of 0–4 [i.e., 0 (no influence), 1
(low influence), 2 (medium influence), 3 (high influence), and 4 (very high influence)]
resulting in non-negative decision matrix of $X_k = [X_{ij}^k]$ with varies the expert(H) from
$1 \leq k \leq H$. At last, initial direct relational matrix [11] is generated by averaging the
non-negative decision matrix of X_k.

$$a_{ij} = \frac{1}{H} \sum_{k=1}^{H} x_{ij}^k \tag{3.3}$$

$$A_{ij} = \begin{bmatrix} a_{11} & a_{12} & \cdots & \cdots & a_{1n} \\ a_{21} & a_{22} & \cdots & \cdots & a_{2n} \\ \cdots & \cdots & \cdots & \cdots & \cdots \\ \cdots & \cdots & \cdots & \cdots & \cdots \\ a_{n1} & a_{n2} & \cdots & \cdots & a_{nn} \end{bmatrix} \tag{3.4}$$

where A_{ij} represents the initial direct relation matrix or average matrix, a_{ij} represents
the average expert opinion values between criteria/output parameters (i) and crite-
ria/output parameters (j), H represents the total number of experts, and k denotes
individual experts.

Step 4: Formulation of a normalized initial direct relation matrix.

The normalized initial direct relation matrix Z (i.e., $Z = [z_{ij}]$) can be obtained by
normalizing the average matrix A_{ij} using the following equations.

$$Z = A/u \tag{3.5}$$

$$u = \max \left(\max_{1 \leq i \leq n} \sum_{j=1}^{n} a_{ij}, \max_{1 \leq j \leq n} \sum_{i=1}^{n} a_{ij} \right) \tag{3.6}$$

Step 5: Computation of total relation matrix

In this step, total relation matrix is determined using Eq. (3.5) based on the values
of normalized initial direct relation matrix obtained using Eq. (3.4). Similarly, the
precise priority weights for each of the criteria/output parameters are determined by
summation of column (i.e., vector r) and summation of rows (i.e., vector c) of matrix
K using Eqs. (3.8) and (3.9), respectively.

$$K = Z + Z^2 + \cdots + Z^m = Z(I - Z)^{-1} \tag{3.7}$$

$$r = [r_i]_{n \times 1} = \left(\sum_{j=1}^{n} k_{ij} \right)_{n \times 1} \tag{3.8}$$

$$c = [c_j]_{1 \times n} = \left(\sum_{j=1}^{n} k_{ij} \right)_{1 \times n} \tag{3.9}$$

where K is the total relation matrix, I is the identity matrix, r_i and c_j denote the sum of rows and columns for each total relation matrix (K), and n is the total no criteria/output parameters.

After that, vector $(r_i + c_i)$ and vector $(r_i - c_i)$ values for each of the criteria/output parameters are calculated to get the precise priority weights (D_j). The elements in vector $(r_i + c_i)$ column values indicate the degree of importance of criterion/output parameter (i) in the system, while vector $(r_i - c_i)$ values represent the net affect which criterion/output parameter (i) makes on the system. If the vector $(r_i - c_i)$ for criterion/output parameter is positive, that indicates the criterion/output parameter is influential while negative vector $(r_i - c_i)$ value of criterion/output parameter signify that the criterion is being influenced by other criteria/output parameters. The summation of row (r_i) and column (c_j) value of criterion/output parameter gives its corresponding DEMATEL weight (D_j).

Step 6: Determination of weighted decision matrix

Then, the weighted decision matrix is determined using Eq. (3.10). It involves the multiplication of DEMATEL weights (D_j) with the performance values of the normalized decision matrix (N_{ij})

$$T = T_{ij} = D_j \times N_{ij} \tag{3.10}$$

where T is the weighted decision matrix, D_j is the DEMATEL weight of the criteria/output parameters.

Step 7: Identification of positive and negative ideal solution

In this step, the positive ideal solution (I^+) and negative ideal solution (I^-) are identified from the weighted decision matrix using Eqs. (3.11) and (3.12). The positive ideal solution indicates the maximum value for beneficiary criteria/output parameters and minimum value for non-beneficiary criteria/output parameters among the various alternatives in the weighted decision matrix. The negative ideal solution indicates the minimum value for beneficiary criteria/output parameters and maximum value for non-beneficiary criteria/output parameters.

$$I^+ = \{V_1^+, V_2^+, \dots V_n^+\}, \quad \text{where } V_j^+ = \{(\max i \ (\text{if } j = J); \ (\min i \ V_{ij} \ \text{if } j = J')\} \tag{3.11}$$

$$I^- = \{V_1^-, V_2^- \dots V_n^-\}, \quad \text{where } V_j^- = \{(\min i \ (V_{ij} \ \text{if } j = J); \ (\max i \ V_{ij} \ \text{if } j = J')\} \tag{3.12}$$

where I^+ and I^- are positive and negative ideal solution matrices, respectively. V_{ij} is the criteria/output parameters' values of the weighted decision matrix with J being associated with beneficial criteria/output parameters and $J^!$ associated with non-beneficial criteria/output parameters.

Step 8: Calculation of Separation Distance for Each of the Criterion

After calculation of I^+ and I^- values for each of the criteria/output parameters, the evaluation of separation distance, i.e., S^+ and S^- from I^+ and I^-, is done using Eqs. (3.13) and (3.14), respectively. Minimum S^+ and maximum S^- signifies that the alternative/input parameter setting is closer to the ideal solution. Similarly, maximum S^+ and minimum S^- signify that the alternative/input parameter is far from the ideal solution.

$$S^+ = \sqrt{\sum_{j=1}^{n}(V_j^+ - V_{ij})^2} \tag{3.13}$$

$$S^- = \sqrt{\sum_{j=1}^{n}(V_j^- - V_{ij})^2} \tag{3.14}$$

where S^+ and S^- are the separation distance of the alternative/input parameters from positive ideal solution (I^+) and negative ideal solution (I^-), respectively, and j is the criterion index.

Step 9: Evaluation of Relative Closeness of Each Alternative

In this step, the closeness values (C_i) are computed using Eq. (3.15) from the separation distance obtained in the previous step. The closeness values (C_i) are evaluated by taking into consideration both beneficiary and non-beneficiary criteria/output parameters, thus converting the multi-objective optimization problem into single-objective problem.

$$C_i = S_i^- / (S_i^+ + S_i^-), \quad 0 \leq C_i \leq 1 \tag{3.15}$$

Step 10: Ranking of the Alternatives

At last, the ranking of alternatives/input parameters' settings or experimental runs settings is done based on the closeness value (C_i) obtained using Eq. (3.15). The alternatives/input parameters' settings or experimental runs with maximum closeness value (Ci) are the most optimal among other alternatives or experimental setting. The optimal combination of process parameters gives the desired output parameters values, thus reducing the manufacturing cost while increasing the productivity and product quality.

3.3 Experimental Study

3.3.1 Background

An experimental investigation on modeling and optimization of AWJM process parameters of process has been selected for validation of the integrated MCDM method. The material used during the machining is sundi wood dust filler-based reinforced polymer (SWDRP) composites [12, 13]. These composites possess special properties like lightweight, high specific strength, small density. They are also highly biodegradable and less expensive. It makes them very useful for auto parts, aerospace, and various other industrial applications.

The sundi wood dust reinforced polymer composite is difficult and expensive to machine by conventional technique, owing to its non-homogeneous and anisotropic properties resulting in filler delamination, filler pulls out, poor surface finish, etc. Hence, AWJM process has been attempted for sundi wood dust reinforced polymer composite machining.

Further, the AWJM process performance strongly depends on its input parameters and output responses. Thus, modeling and optimization of AWJM process on machining of sundi wood dust reinforced polymer composites is essential. Therefore, in the present chapter, an attempt has been made to use integrated MCDM method, i.e., M-DEMATEL with TOPSIS method, for modeling and optimization of AWJM process. In this, DEMATEL is used for determining the criteria weights and TOPSIS for identification of optimum AWJM process parameters. Finally, the ideal grouping of the input parameters for practical applications is suggested. A recommendation is made for optimal process parameters combination. The identified optimal parameter setting reduces the cost and improves the quality of the product and performance of the AWJM process while machining of sundi wood dust reinforced polymer composite.

3.3.2 Workpiece Preparation

Sundi wood dust (SWD) as reinforcement material with a particle size of 400 μm and density of 0.779 gm/cc is used for preparation of work specimen. The main constituents of the specimen are cellulose, glucomannan, xylem, and linen. The sundi wood dust and the working specimen are shown in Fig. 3.2a–b.

The percentage of filler used in the composite is 3% of the total composition, and the rest is a matrix (97%) comprising epoxy (grade LY 556) with density 1.26 gm/cc and hardener (HY 951). The resin and hardener are mixed in the ratio of 10:8 by weight. The reinforcement material sundi wood dust and the matrix are then mechanically stirred and poured into the vacuum glass chamber and allowed to cure for a minimum period of 24 h at ambient temperature. The specimen having dimension 180 mm × 140 mm × 6 mm is then taken for actual machining.

Fig. 3.2 a Sundi wood dust. **b** Prepared specimen

3.3.3 Experimental Procedure

The AWJM machine manufactured by DARDI International Corporation, China, as seen in Fig. 3.3a–b is used for machining of sundi wood dust. In the present work, four parameters such as a standoff distance (SoD), working pressure (WP), nozzle speed (NS), and abrasive grain size (AGS) are experimented with three levels (Table 3.1), and Taguchi (L9) orthogonal array is utilized to design experiments. During experimentation, working pressure of 3800 bar, the discharge rate of 2.31 l/min, an orifice diameter of 0.25 mm, and abrasive material with a mesh size of 70, 80, and 90 microns are used [12]. Throughout the experiments, the voltage of 300 V, current of 20 A, and nozzle angle of 90° are set in the AWJM machine. In the experimentation, 20 mm × 20 mm square holes are made using the AWJM machine. Each experiment was per-

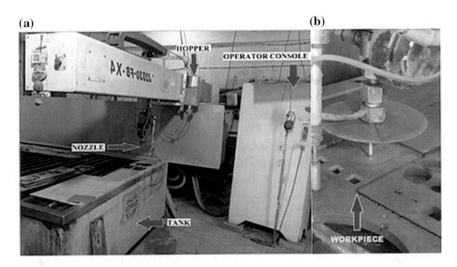

Fig. 3.3 a AWJM experimental setup. **b** AWJM nozzle head setup

Table 3.1 Input parameters and their levels for AWJM process

Input parameters	Symbol	Units	Level 1	Level 2	Level 3
Standoff distance	SoD	mm	1	2	3
Work pressure	WP	MPa	100	125	150
Nozzle speed	NS	mm/min	100	200	300
Abrasive grain size	AGS	mesh	70	80	90

Table 3.2 Experimental combinations of process parameters and corresponding values of responses/output parameters

Exp. no.	Input parameters				Output parameters		
	SoD (mm)	WP (MPa)	NS (mm/min)	AGS (mesh)	MRR (g/s)	SR (μm)	TA (°)
1	1	100	100	70	3.251	0.15	0.189
2	1	125	200	80	12.652	0.143	0.406
3	1	150	300	90	28.927	0.172	0.301
4	2	100	200	90	24.956	0.13	0.426
5	2	125	300	70	47.853	0.135	0.556
6	2	150	100	80	18.6251	0.185	0.279
7	3	100	300	80	66.771	0.16	0.118
8	3	125	100	90	24.784	0.102	0.119
9	3	150	200	70	55.551	0.178	0.234

formed three times, and their average values are taken for material removal rate (MRR), surface roughness (SR), and taper angle (TA) in the analysis (Table 3.2). During experimentation, output parameter [i.e., material removal rate(MRR) and taper angle (TA)] are evaluated using the following equations:

$$MRR = \frac{W_a - W_b}{t} \tag{3.15}$$

$$TA(\theta) = \tan^{-1}\frac{K_t - K_b}{2t} \tag{3.16}$$

where W_a and W_b are initial and final weight, respectively; t is the cutting time; t_w is the thickness of workpiece, K_t and K_b are the top and bottom kerf width, respectively.

3.3.4 Modeling of AWJM Process

The modeling of AWJM process on machining of sundi wood dust reinforced polymer composites is done using M-DEMATEL–TOPSIS method. First, decision-making matrix has been developed based on Taguchi (L9) orthogonal array design, and the results of decision matrix are shown in Table 3.2. The decision matrix includes a num-

Table 3.3 Normalized data of AWJM response parameters

Exp. no.	MRR (g/mm)	SR (μm)	TA (°)
1	0.0292	0.3277	0.1947
2	0.1135	0.3124	0.4182
3	0.2595	0.3757	0.3101
4	0.2239	0.2840	0.4388
5	0.4293	0.2949	0.5727
6	0.1671	0.4041	0.2874
7	0.5990	0.3495	0.1215
8	0.2223	0.2223	0.1226
9	0.4983	0.3888	0.2410

Table 3.4 Average expert opinion matrix

Output parameters	MRR (g/mm)	SR (μm)	TA (°)
MRR (g/mm)	0	3	1
SR (μm)	0.3333	0	0
TA (°)	1	0	0

ber of criteria [i.e., a number of output parameters like material removal rate (MRR), surface roughness (SR), and taper angle (TA)] and its corresponding alternatives (i.e., the number of experimental runs having optimal input/process parameters setting). The criteria material removal rate (MRR) is considered to be beneficial criteria or output parameter, and higher values are desirable while the response parameters like surface roughness (SR) and taper angle (TA) are considered to be non-beneficial criteria/output parameters and lower values are desirable. Second, normalization of decision-making matrix is performed to convert original sequence of data of output parameters into comparable or non-dimensional data using Eq. (3.2). The results of normalized decision-making matrix are shown in Table 3.3.

Third, priority weights for each of the criteria or output parameters are computed using M-DEMATEL method. In the M-DEMATEL method, first, construction of comparative decision matrix between the criteria, i.e., output parameters based on the expert's opinion using a scale of 0–4, is done. Four members of experts are considered for the study, and each of their performance values with respect to each of the criteria, i.e., output parameters, are collected. Then, average values (Table 3.4) of each of the expert opinion against each of the output parameters are taken for analysis. Second, normalization of expert opinion matrix is carried out using Eqs. (3.5) and (3.6), respectively, and results are illustrated in Table 3.5. Third, total relation matrix and precise priority weights for each of the output parameters are determined using Eqs. (3.7)–(3.9), respectively. The results of total relation matrix and corresponding precise priority weights for each of output parameters are shown in Table 3.6.

Table 3.5 Normalized direct relation matrix

Output parameters	MRR	SR	TA
MRR	0	0.75	0.25
SR	0.0833	0	0
TA	0.25	0	0

Table 3.6 Total relationship matrix and priority weights

Output parameters	MRR	SR	TA	r_i	c_i	$r_i - c_i$	$r_i + c_i = D_j$
MRR	0.1429	0.8571	0.2857	1.2857	0.5238	0.7619	1.8095
SR	0.0952	0.0714	0.0238	0.19047	1.1428	−0.9524	1.3333
TA	0.2857	0.2143	0.0714	0.57143	0.3809	0.1905	0.9524

Table 3.7 Weighted decision matrix

Exp. no.	MRR (g/mm)	SR (μm)	TA (°)
1	0.0528	0.4369	0.1854
2	0.2054	0.4165	0.3983
3	0.4696	0.5010	0.2953
4	0.4051	0.3786	0.4179
5	0.7768	0.3932	0.5455
6	0.3023	0.5388	0.2737
7	1.0839	0.4660	0.1158
8	0.4023	0.2971	0.1167
9	0.9018	0.5184	0.2296

It can be observed from Table 3.6 that the $r_i + c_i$ values for output responses like material removal rate (MRR), surface roughness (SR), and taper angle (TA) are identified as 1.8095, 1.3333, and 0.9524, respectively. The high $r_i + c_i$ value of the output parameter, i.e., material removal rate (1.8095), indicates that it is the most influential parameter while the low $r_i + c_i$ value of the output parameter, i.e., taper angle (0.9524), is an indication that it is least influential parameter. Similarly, the values of $(r_i - c_i)$ for each of the output parameters like material removal rate (MRR), surface roughness (SR), and taper angle (TA) are determined to be 0.7619, −0.9524, and 0.1905, respectively. It is seen that material removal rate (MRR) found to be of more influence on the outcome of other criteria or output parameter due to the positive $(r_i - c_i)$ value. On the other side, surface roughness (SR) is found to be negative $(r_i - c_i)$ value which indicates that the criterion gets more affected due to changes in other criteria.

Thereafter, weighted decision matrix (Table 3.7) is derived based on the precise priority weights using Eq. (3.10). The weighted decision matrix is obtained by multiplying the precise priority weights obtained from DEMATEL method with the performance values of initial normalized decision matrix.

Table 3.8 Positive and negative ideal solution values

Output parameters	V_j^+	V_j^-
MRR	1.0839	0.0528
SR	0.2971	0.5388
TA	0.1158	0.5455

Table 3.9 Separation distance of alternatives from PIS and NIS

Exp. no.	S^+	S^-
1	0.8217	0.4620
2	0.4766	0.1169
3	0.2309	0.1288
4	0.2951	0.0646
5	0.2187	0.5784
6	0.3819	0.0222
7	0.1689	0.5286
8	0.6806	0.3209
9	0.1530	0.5127

Then, the positive ideal solution (I^+) and negative ideal solution (I^-) for each of the output parameters are computed by Eqs. (3.11) and (3.12), respectively, which are depicted in Table 3.8. The positive ideal solution for the output parameter, i.e., material removal rate (MRR), is taken to be larger the better; while the output parameters like surface roughness (SR) and taper angle (TA) are taken to be smaller the better. Similarly, negative ideal solution for the output parameter, i.e., material removal rate (MRR), is taken as smaller the better; while response parameters like surface roughness (SR) and taper angle (TA) are taken as larger the better. After that, the separation distances (S^+ and S^-) of the alternatives from positive (I^+) and negative ideal solution (I^-) are computed using Eqs. (3.13) and (3.14), respectively, and whose results are depicted in Table 3.9. The results show that the separation distance of experiment number 9 from positive ideal solution (I^+) is the least with 0.1689 and is the most from negative ideal solution (I^-) with 0.5286.

At last, the closeness coefficient (C_j) values of each alternatives/input process parameters settings are calculated using Eq. (3.15). This converts the multi-objective problem into a single-objective problem [14, 15]. The closeness coefficient values (C_j) for each of the alternatives/experimental runs or input parameter settings are illustrated in Table 3.10. Subsequently, the experimental runs are ranked based on the closeness coefficient (C_j) values obtained from the proposed M-DEMATEL–TOPSIS method.

Table 3.10 Data of relative closeness

Exp. no.	Input parameters				Closeness coefficient (C_j) values
	SoD (mm)	WP (MPa)	NS (mm/min)	AGS (mesh)	
1	1	100	100	70	0.3599
2	1	125	200	80	0.1969
3	1	150	300	90	0.3580
4	2	100	200	90	0.1796
5	2	125	300	70	0.7256
6	2	150	100	80	0.0549
7	3	100	300	80	0.7578
8	3	125	100	90	0.3204
9	3	150	200	70	0.7702

3.4 Results and Discussion

3.4.1 Optimum Combination of Process Parameters

Optimization of AWJM process parameters is done based on the relative closeness coefficient (C_j) values of each of the experimental runs. M-DEMATEL with TOPSIS method is used for AWJM process modeling and optimization while machining of sundi wood dust reinforced polymer composites. The optimized values of abrasive water jet machining process are shown in Table 3.11. Subsequently, based on the relative closeness value, the ranking of the alternative/experimental run or input parameter setting is carried out from the proposed DEMATEL–TOPSIS technique.

Table 3.11 Closeness coefficient and rank corresponding to each experimental combination in AWJM process

Exp. no.	Input parameters				Closeness coefficient (C_j) values	Rank
	SoD (mm)	WP (MPa)	NS (mm/min)	AGS (mesh)		
1	1	100	100	70	0.3599	4
2	1	125	200	80	0.1969	7
3	1	150	300	90	0.358	5
4	2	100	200	90	0.1796	8
5	2	125	300	70	0.7256	3
6	2	150	100	80	0.0549	9
7	3	100	300	80	0.7578	2
8	3	125	100	90	0.3204	6
9	3	150	200	70	0.7702	1

It can be observed that the closeness coefficient (C_j) value (Table 3.11) of experiment number 9 is the highest with 0.7702, and hence, the alternative/experimental run or input parameter setting corresponding to experiment number 9 is the most optimal setting among the nine experimental runs.

The optimal combination of AWJM parameters is standoff distance (SoD)—3 mm, working pressure (WP)—150 MPa, nozzle speed (NS)—200 mm/min, and abrasive grain size (AGS)—70 mesh. The M-DEMATEL–TOPSIS helped to obtain the optimal setting in abrasive water jet machining process. Furthermore, optimal setting reduces the process time, reduces the waste and pollution, increases the rate of material removal rate, and provides better surface finish and less taper angle. In other words, it improves quality and machining performance and reduces manufacturing cost. Overall, the investigation recommends integrated MCDM method, i.e., M-DEMATEL–TOPSIS method, for modeling and optimization of AWJM process for high-performance machining of reinforced polymer composites.

3.4.2 Optimum Input Parameters and Level Combination

The determination of optimum input parameters and their level combination are carried out by averaging the closeness coefficient (C_j) value for each level of input parameters, and the corresponding results are depicted in Table 3.12. As the closeness coefficient (C_j) value represents the level of correlation between the reference and comparability sequence. Hence, comparability sequence has higher closeness coefficient (C_j) value for optimum parameters which are standoff distance (SoD) at level 3, working pressure (WP) at level 1, nozzle speed (NS) at level 3, and abrasive grain size (AGS) at level 1.

Furthermore, most significant input parameters are determined by calculating the difference between maximum and minimum values of the mean closeness coefficient (C_j) value. The mean closeness coefficient (C_j) values for each of the input parameters obtained are standoff distance (SoD) as 0.311, working pressure (WP) as 0.038, nozzle speed (NS) as 0.369, and abrasive grain size (AGS) as 0.333. It is seen that parameter nozzle speed (NS) is found to be more significant input parameter while

Table 3.12 Average closeness coefficient (C_j) value

Input parameters	Average closeness coefficient (C_j) value			Max − Min	Rank
	Level 1	Level 2	Level 3		
SoD	0.305	0.320	0.616[a]	0.311	3
WP	0.432[a]	0.414	0.394	0.038	4
NS	0.245	0.382	0.614[a]	0.369	1
AGS	0.619[a]	0.337	0.286	0.333	2

[a]Optimal level

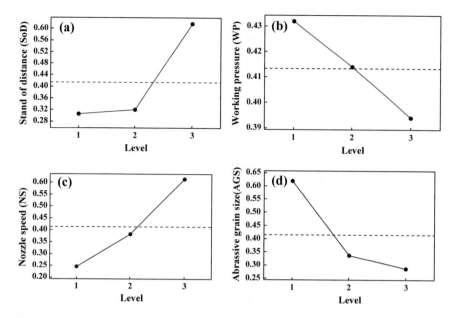

Fig. 3.4 a–d Response graphs of closeness coefficient (C_j) value

working pressure (WP) as least significant parameters for AWJM process. The order of importance of the input parameters for AWJM process can be listed as nozzle speed (NS)–abrasive grain size (AGS)—standoff distance (SoD)—working pressure (WP).

Additionally, the influence of each input parameter is represented by response graphs as shown in Fig. 3.4. The graphical representation shows that, there is a negligible changes in the closeness coefficient (Cj) value when input parameters like SoD, WP, NS and AGS varies their level from 1 to 3. However, maximum closeness coefficient (Cj) value for AWJM process is obtained at optimum conditions of the input parameters.

3.5   Summary

This chapter reports the modeling and optimization of AWJM process using TOP-SIS–DEMATEL integrated multi-criteria decision-making technique. To show the strength and applicability of the proposed method, experimental cases study on optimization of AWJM process on machining of sundi wood dust reinforced polymer composites are presented. From the study, it is concluded that machining of sundi wood dust reinforced polymer composites using abrasive water jet machining is feasible, and the use of M-DEMATEL–TOPSIS methodology yields the optimal solution. The result reveals that experiment number 9 leads to the optimal parameter combinations and provides the desirable output parameters for AWJM process.

The optimal parameters attained are 3 mm SoD, WP of 150 MPa, 200 mm/min NS, and 70 mesh abrasive grain size. The optimal setting gives higher MRR, better surface quality, and lesser taper angle while machining of sundi wood dust reinforced polymer composites using AWJM process. In addition, optimal setting obtained improves the AWJM process efficiency through better product quality with less generation of waste, pollution, and better productivity at low cost. Thus, it is concluded that the proposed integrated method shall be utilized as a competent model for modeling and optimization of various advanced machining processes.

References

1. Gary FB (1987) Non-traditional manufacturing processes. Marcel Dekker, New York
2. Raval MA, Patel CP (2013) Parametric optimization of magnetic abrasive water jet machining Of AISI 52100 steel using grey relational analysis. Int J Eng Res Appl (IJERA) 3(4):527–530
3. Azmir MA, Ahsan AK, Rahmah A, Noor MM, Aziz AA (2007) Optimization of abrasive water jet machining process parameters using orthogonal array with grey relational analysis. Regional Conference on engineering mathematics, mechanics, manufacturing & architecture, pp 21–30
4. Gaidhani YB, Kalamani VS (2013) Abrasive water jet review and parameter selection by AHP method. IOSR J Mech Civil Eng 8(5):1–6
5. Jegaraj JJR, Babu NR (2007) A soft computing approach for controlling the quality of cut with abrasive water jet cutting system experiencing orifice and focusing tubewear. J Mater Process Technol 185(1–3):217–227
6. Iqbal A, Dar NU, Hussain GJ (2011) Optimization of abrasive water jet cutting of ductile materials Wuhan Univ Technol-Mat Sci Edit 26(1):88–92
7. Chaturvedi V, Singh D (2015) Multi response optimization of process parameters of abrasive water jet machining for stainless steel AISI 304 using VIKOR approach coupled with signal to noise ratio methodology. J Adv Manuf Syst 14:107–121
8. Gabus A, Fontela E (1972) World problems an invitation to further thought within the framework of DEMATEL. Battelle Geneva Research Centre, Switzerland Geneva
9. Gabus A, Fontela E (1973) Perceptions of the world Problematique: communication procedure, communicating with those bearing collective responsibility (DEMATEL Report No. 1). Battelle Geneva Research Centre, Switzerland, Geneva
10. Hwang CL, Yoon K (1981) Multiple attribute decision making: methods and applications. Springer, New York
11. Lin CJ, Wu WW (2008) A causal analytical method for group decision making under fuzzy environment. Expert Syst Appl 34(1):205–213
12. Kumar R, Kumar K, Sahoo P, Bhowmik S (2014) Study of mechanical properties of wood dust reinforced epoxy composite. Procedia materials science 6:551–556
13. Pritchard G (2004) Two technologies merge wood-plastic composites. Plast Addit Compouding 6:18–21
14. Brauers WKM, Zavadskas EK (2006) The MOORA method and its application to privatization in a transition economy. Control Cybern 35(2):445–469
15. Prasad K, Chakraborty S (2012) Application of multi-objective optimization on the basis of ratio analysis (MOORA) method for materials selection. Mat Des 37:317–324

Chapter 4
Modeling and Optimization of Ultrasonic Machining Process

4.1 Introduction

Ultrasonic machining process (USM) is a non-traditional machining process used for machining of hard and brittle materials such as tungsten carbide, diamond, glass, silicon quartz, and difficult-to-machine materials, i.e., titanium. [1]. In ultrasonic machining process, the material removal takes place by micro-erosion caused due to impact of abrasive particles held between tool and workpiece in the form of slurry under the influence of high-frequency vibrations [1–3]. The major components of ultrasonic machining include transducer, horn, feed mechanism, slurry pump, and slurry tank. Figure 4.1 illustrates various parts of ultrasonic machining system and working principle of ultrasonic machining process. The various process parameters include amplitude of vibration, frequency of vibration, feed force, abrasive size, and abrasive material which affect the performance of ultrasonic machining process.

Since ultrasonic machining process is characterized by low material removal rate [1–4], it is therefore extremely important to make suitable steps for improvement of metal removal rate without affecting the surface quality of the work surface. Optimization of machining parameters is the prominent approach to improve MRR.

Selecting the process parameters through operator experience or handbooks may not lead to optimal output responses which are complicated in nature as we need to maximize material removal rate (MRR) while minimizing surface roughness (SR), tool wear rate (TWR), taper angle (TA), and overcut (OC). Moreover, the process parameters with optimization may not yield the desired job characteristics which subsequently affect the performance, efficiency, tool life of the machining process, and quality of the finished product. Therefore, a suitable multi-criteria decision-making technique (MCDM) is required to be implemented for modeling and optimization of the ultrasonic machining process parameters.

Some investigations have been conducted on modeling and optimization of ultrasonic machining process parameters using MCDM techniques such as adaptive neuro-fuzzy inference system [5], principal component analysis (PCA)-based grey

Fig. 4.1 Working principle
of ultrasonic machining
process

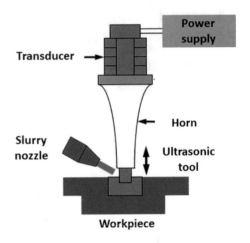

relational analysis (GRA) and principal component (PCA)-based TOPSIS [6], and
other optimization approaches such as genetic algorithm (GA) [7], gravitational
search and fireworks algorithms [8], Taguchi approach [9, 10], artificial bee colony
(ABC), particle swarm optimization (PSO) and harmony search (HS) algorithms
[11]. However, it can be observed that very few articles are available on modeling and
optimization of ultrasonic machining process parameters using MCDM techniques.
Similarly, existing approaches have not been that much successful to consider the
priority weights for each of the objective and subjective parameters during the optimization of ultrasonic machining process. Furthermore, consideration of beneficial
and non-beneficial parameters and their effect during the selection of optimal setting
for ultrasonic machining process have been ignored [12]. Besides, existing methods
do not consider the uncertain, imprecise, and vagueness information of output parameters during modeling and optimization. At last, existing methods include complex
steps and lengthy procedure to solve the multiple response parameter optimization
problems. Considering the drawbacks of these methods, in the present chapter, an
effort is made to verify the integrated MCDM method for modeling and optimizing
the USM process parameter data. The integrated MCDM method consists of fuzzy
with multi-criteria ratio analysis (F-MCRA) methods. Here, fuzzy method determines the precise priority weights for each of the objective and subjective parameters during optimization of USM process, while multi-criteria ratio analysis has
been used for selection of optimal process parameters considering the effect of both
beneficial and non-beneficial parameters in account.

4.2   Integrated Fuzzy-MCRA-Based MCDM Method

The proposed method consists of fuzzy theory (Lotfi A. Zadeh, 1965) combined with multi-criteria ranking analysis (MCRA). The detailed steps of fuzzy-MCRA-based decision-making method are as follows,

Step 1: Development of Decision Matrix

First, the decision matrix is developed which consists of criteria as output/response parameters and corresponding alternatives as process parameters settings. The decision matrix comprises performance values of the output responses obtained by varying the process parameters as per the design of experiment. The decision matrix is formulated using Eq. 4.1.

$$D_{ij} = \begin{bmatrix} & C1 & C2 & \ldots & Cn \\ A1 & Y_{11} & Y_{12} & \ldots & Y_{1n} \\ A2 & Y_{21} & Y_{22} & \ldots & Y_{2n} \\ \ldots & \ldots & \ldots & \ldots & \ldots \\ Am & Y_{m1} & Y_{m2} & \ldots & Y_{mn} \end{bmatrix} \tag{4.1}$$

where D_{ij} is the decision matrix, $C_1, C_2 \ldots C_n$ represent the number of criteria/output parameters, $A_1, A_2 \ldots A_n$ indicate the number of alternatives/number of experimental runs/process parameters, and a_{ij} indicates the performance values of the output responses corresponding to each of the experimental settings.

Step 2: Normalization of Decision Matrix

Normalization of the decision matrix is done to convert the non-comparable criteria/output parameters with different units into dimensionless comparable criteria/output parameters with value's ranging from 0 to 1. The normalized matrix depicts the relative performance of the alternatives with respect to the criteria. Normalization of the decision matrix is done using the following equations.

$$N_{ij} = \frac{Y_{ij}}{\max(Y_{ij})} \quad \text{For beneficial criteria} \tag{4.2}$$

$$N_{ij} = \frac{\min(Y_{ij})}{Y_{ij}} \quad \text{For non-beneficial criteria} \tag{4.3}$$

where N_{ij} represents the normalized decision matrix, $\max(Y_{ij})$ indicates the maximum performance value of the criteria/output parameters for beneficial criteria, and $\min(Y_{ij})$ is the minimum performance value criteria/output parameters for non-beneficial criteria.

Table 4.1 Linguistic variables and TFN [13]

Importance	Abbreviation	TFN
Extremely low	EL	$(0, 0, 0.1)$
Very low	VL	$(0, 0.1, 0.3)$
Low	L	$(0.1, 0.3, 0.5)$
Medium	M	$(0.3, 0.5, 0.7)$
High	H	$(0.5, 0.7, 0.9)$
Very high	VH	$(0.7, 0.9, 1)$
Extremely high	EH	$(0.9, 1, 1)$

Step 3: Identification of Criteria Weights

In this step, criteria/output parameters weights are identified via linguistic variables. The linguistic variables for each of the criteria/output parameters are assigned by experts based on their importance. Then, assigned linguistic variables for each of the criteria/output parameters are converted into fuzzy weights via aggregation of triangular fuzzy numbers (TFNs). The linguistic variables and their corresponding triangular fuzzy number are shown in Table 4.1.

Step 4: Formulation of Weighted Decision Matrix

The priority weights for each of the criteria/output parameters obtained in the previous step are multiplied with the normalized decision matrix using Eq. (4.4) to obtain the weighted decision matrix (WDM). The weighted decision matrix gives the relative weights of each criteria/output parameters with respect to the alternatives/process parameters settings.

$$\text{WDM} = N_{ij} \times W_j \tag{4.4}$$

where WDM is the weighted decision matrix, W_j denotes the priority weights for each of the criteria/output parameters, and N_{ij} represents the normalized decision matrix.

Step 5: Evaluation of Closeness Index

In the final step, closeness index (CI) is obtained by subtracting the sum of non-beneficiary criteria (NBC) value from the sum of beneficiary criteria (SBC) value for each alternative from the weighted decision matrix using following expressions.

$$\text{SBC} = \sum_{i=1}^{b} \text{WDM} \tag{4.5}$$

$$\text{NBC} = \sum_{i=b+1}^{n} \text{WDM} \tag{4.6}$$

$$\text{CI} = \text{SBC} - \text{NBC} \tag{4.7}$$

where SBC and NBC are the sum of beneficial and non-beneficial criteria/output parameters' values of weighted decision matrix, respectively, b denotes the number of beneficiary criteria/output parameters, n represents the total number of criteria/output parameters, and CI is the closeness index.

Step 6: Ranking of the Alternatives

At last, ranking of the alternatives is done based on the CI values. The alternative with the maximum positive CI value or minimum negative CI value provides the optimal process parameter setting among the other alternatives/process parameter settings.

4.3 Experimental Study

4.3.1 Background

The present experimental study is about validation of the integrated MCDM method (i.e., fuzzy-MCRA) for modeling and optimization of USM process parameters. The material used during the machining is Zirconia (ZrO_2) composite. Zirconia is basically a ceramic material used in several industrial applications due to its superior properties and unique advantages such as higher fracture toughness, excellent wear resistance, and high-temperature stability [10]. Machining of Zirconia (ZrO_2) composite using conventional machining is a difficult process as it is a hard and brittle material. Hence, non-conventional machining process, namely USM, is employed in the present chapter as it is an effective process for machining of ceramics with less physical, chemical, and metallurgical property changes.

4.3.2 Workpiece Preparation

The work specimen of Zirconia (ZrO_2) composite is prepared by sol–gel method whose bulk density was 6 g/cc, and it stabilized with 8 mol% of yttria (yttrium oxide, Y_2O_3). The specimen size considered for experimentation is 100 mm × 100 mm × 3 mm. The Zirconia (ZrO_2) composite specimen is shown in Fig. 4.2.

4.3.3 Experimental Procedure

The Zirconia (ZrO_2) composite is machined using sonic mill AP-1000 ultrasonic machining (Fig. 4.3). Three process parameters are considered such as slurry concentration (SC), feed rate (FR), and power (P) with three levels each (Table 4.2).

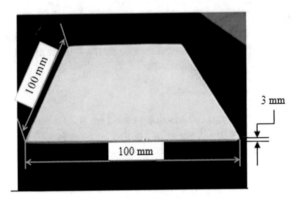

Fig. 4.2 Ceramic, i.e., Zirconia (ZrO$_2$) composite, specimen

Fig. 4.3 AP-1000 ultrasonic machine used for machining of Zirconia

Taguchi (L9) orthogonal array is utilized for design of experiments [14]. Stainless steel ultrasonic machining tool with 1.61 mm cross section is used for machining, and the abrasive used is boron carbide (B$_4$C) with grit size of 400. For better accuracy, each experiment is repeated three times and their average is taken for further analysis. The output parameters such as material removal rate (MRR), taper angle (TA), and overcut (OC) are calculated using Eqs. (4.8–4.10).

Table 4.2 Ultrasonic machining process parameters and levels

Input parameter	Unit	Symbol	Levels		
			1	2	3
Slurry concentration	%	SC	40	50	60
Feed rate	mm/min	FR	0.90	1.15	1.40
Power	Watt	P	300	350	400

$$\text{MRR} = \frac{\pi R^2 h}{t} \tag{4.8}$$

$$\text{TA} = \tan^{-1} \frac{R_{\text{entry}} - R_{\text{exit}}}{h} \tag{4.9}$$

$$\text{OC} = R_{\text{entry/exit}} - R_{\text{tool}} \tag{4.10}$$

where t denotes the machining time, h represents the height of the hole (plate thickness), R is the radius of the hole, R_{tool} is the radius of the tool. The experimental results based on the design of experiments are shown in Table 4.3.

4.3.4 Modeling of USM Process

This section is about modeling and optimization of USM process parameters on machining of Zirconia (ZrO$_2$) composite using the developed integrated MCDM method, i.e., fuzzy-MCRA. First, the decision matrix is developed based on Taguchi

Table 4.3 Taguchi-based experimental design and results

Exp. no.	Input parameters			Output parameters			
	SC (%)	FR (mm/min)	P (watt)	MRR (g/mm)	TA (°)	OC (μm)	
						Entry	Exit
1	40	0.90	300	1.169	0.458	44.5	20.5
2	40	1.15	350	1.38	0.449	42.5	19.0
3	40	1.40	400	1.726	0.439	41.0	18.0
4	50	0.90	350	1.226	0.42	40.5	18.5
5	50	1.15	400	1.314	0.401	38.0	17.0
6	50	1.40	300	1.651	0.372	35.0	15.5
7	60	0.9	400	1.195	0.353	34.5	16.0
8	60	1.15	300	1.397	0.344	33.0	15.0
9	60	1.40	350	1.602	0.334	32.0	14.5

(L9) orthogonal array (Table 4.3). The decision matrix comprises number of criteria [i.e., the output parameters like material removal rate (MRR), taper angle (TA), and overcut (OC)] and alternatives (i.e., the number of experimental runs which includes the optimal process parameter setting). In the present chapter, the output parameter, i.e., material removal rate (MRR), is considered as beneficial criteria/output parameters, while output parameters, taper angle (TA) and overcut (OC), are considered as non-beneficial criteria/output parameters. Second, the normalization of the decision matrix is done to convert the criteria/output parameters with different units into comparable units using Eqs. (4.2) and (4.3). The normalized results are presented in Table 4.4.

Third, the weights of the criteria/output parameters are calculated by assigning linguistic variables. The linguistic variables assigned by different experts are converted into their corresponding triangular fuzzy numbers and averaged (Tables 4.5 and 4.6). The highest fuzzy weight shows most influencing parameter, while least value of fuzzy weight shows least influencing parameter. It can be observed that parameter, i.e., material removal rate (MRR), yields the highest fuzzy weight with 0.6083 value, and overcut is the least fuzzy weight with 0.4750 value.

Next, the weighted decision matrix is developed by multiplying the fuzzy weights for each of the criteria/output parameters with the normalized decision matrix using Eq. (4.4). The results of weighted decision matrix are shown in Table 4.7.

At last, the closeness index (CI) of each alternatives/process parameters is obtained by subtracting the sum of non-beneficiary criteria/output parameters' values from the sum of beneficiary criteria/output parameters' values. The alternative with highest

Table 4.4 Normalized data of ultrasonic machining response parameters

Exp. no.	MRR (g/mm)	TA (°)	OC (μm)	
			Entry	Exit
1	0.6773	0.7293	0.7191	0.7073
2	0.7995	0.7439	0.7529	0.7632
3	1.0000	0.7608	0.7805	0.8056
4	0.7103	0.7952	0.7901	0.7838
5	0.7613	0.8329	0.8421	0.8529
6	0.9565	0.8978	0.9143	0.9355
7	0.6924	0.9462	0.9275	0.9062
8	0.8094	0.9709	0.9697	0.9667
9	0.9282	1.0000	1.0000	1.0000

Table 4.5 Fuzzy weight matrix for the output responses

Output response	Triangular fuzzy weights		
MRR (g/mm)	0.6333	0.8333	0.9667
TA (°)	0.5667	0.7667	0.9333
OC (μm)	0.4333	0.6333	0.8333

Table 4.6 Priority weights of ultrasonic machining response parameters

Criteria		Fuzzy weights
MRR (g/mm)		0.6083
TA (°)		0.5667
OC (μm)	Entry	0.4750
	Exit	0.4750

Table 4.7 Weighted decision matrix

Exp. no.	MRR (g/mm)	TA (°)	OC (μm)	
			Entry	Exit
1	0.4120	0.4133	0.3416	0.3360
2	0.4864	0.4215	0.3576	0.3625
3	0.6083	0.4311	0.3707	0.3826
4	0.4321	0.4506	0.3753	0.3723
5	0.4631	0.4720	0.4000	0.4051
6	0.5819	0.5088	0.4343	0.4443
7	0.4212	0.5362	0.4406	0.4304
8	0.4924	0.5502	0.4606	0.4591
9	0.5646	0.5667	0.4750	0.4750

positive closeness index or least negative closeness index is the most optimum process parameter setting. It can be observed (Table 4.8) that the least negative closeness index (CI) value of −0.5761 is obtained corresponding to experiment number 3. Hence, the parameter combination at experiment number 3 is the optimal setting among the nine experiments.

Table 4.8 Closeness index values

Exp. no.	Input parameters			Closeness index (CI) values
	SC (%)	FR (mm/min)	P (watt)	
1	40	0.9	300	−0.6788
2	40	1.15	350	−0.6553
3	**40**	**1.4**	**400**	**−0.5761**
4	50	0.9	350	−0.7661
5	50	1.15	400	−0.814
6	50	1.4	300	−0.8055
7	60	0.9	400	−0.986
8	60	1.15	300	−0.9776
9	60	1.4	350	−0.952

4.4 Results and Discussion

4.4.1 Optimum Combination of Process Parameters

The closeness index (CI) values of each of the alternatives or experimental runs or process parameter settings are used to optimize the process parameters of USM process. The optimization results of USM process performed using fuzzy-MCRA method are shown in Table 4.9. Ranking of the USM process is carried out based on the closeness index (CI) values obtained from the fuzzy-MCRA method.

The optimization results show that experiment number 3 is the optimal experimental combination among the nine alternatives or experimental runs or process parameter settings. The closeness index (CI) value for the alternative or experimental run or process parameter setting is −0.5761 and is the least negative value among the nine alternatives or experimental runs.

The optimal settings obtained are slurry concentration (SC) as 40% with level 1, feed rate (FR) as 1.4 mm/min with level 3, and power (P) as 400 W with level 3. The optimal setting obtained using fuzzy-MCRA technique provides the most optimal response parameters for USM process. The optimal setting reduces the machining time and production cost due to high material rate (MRR) and low wastage. Furthermore, the recommended optimal alternative or experimental run or process parameter setting enhances the product quality, efficiency, and performance of the ultrasonic machining process. From the present research, it is observed that the integrated method, i.e., fuzzy-MCRA method, can be employed as an effective decision-making model for modeling and multi-criteria optimization of USM process.

Table 4.9 Experimental runs ranking in optimization of USM process

Exp. no.	Process parameters			CI Values	Rank
	SC (%)	FR (mm/min)	P (Watt)		
1	40	0.9	300	−0.6788	3
2	40	1.15	350	−0.6553	2
3	**40**	**1.4**	**400**	**−0.5761**	**1**
4	50	0.9	350	−0.7661	4
5	50	1.15	400	−0.8140	6
6	50	1.4	300	−0.8055	5
7	60	0.9	400	−0.9860	9
8	60	1.15	300	−0.9776	8
9	60	1.4	350	−0.9520	7

Table 4.10 Average closeness index (CI) values

Input parameters	Average closeness index (CI) values			Max − Min	Rank
	Level 1	Level 2	Level 3		
SC	0.637	0.795	0.972[a]	0.335	1
FR	0.810	0.816[a]	0.778	0.038	2
P	0.821[a]	0.791	0.792	0.030	3

[a]Optimal level

4.4.2 Optimum Input Parameters and Level Combination

This chapter also analyzes the response of input parameters of USM process on the closeness index (CI) values via averaging the closeness index (CI) values for each level of input parameters such as slurry concentration (SC), feed rate (FR), power (P). The results of analysis are shown in Table 4.10. The result shows that closeness index (CI) values for each of the input parameters show better correlation between the reference and comparability sequence. Hence, comparability sequence has closeness index (CI) values for optimum parameters are slurry concentration (SC) at level 3, feed rate (FR) at level 2, and power (P) at level 1.

Furthermore, most significant input parameters are determined by calculating the difference between maximum and minimum values of the mean closeness index (CI) values. The mean closeness index (CI) values for each of the input parameters obtained are slurry concentration (SC) 0.335, feed rate (FR) 0.038, and power (P) 0.030. It is observed from the results that parameter slurry concentration (SC) found to be more significant input parameters while power (P) as least significant parameters for USM process. The order of importance of the input parameters for USM process can be listed as slurry concentration (SC)—feed rate (FR)—power (P).

Additionally, the influence of each input parameter can be seen in the graphs as shown in Fig. 4.4. The graphs show that, small variation in closeness index (CI) values for the USM process when input parameters such as SC, FR and P varies from level 1 to level 3. Further, maximum output for USM during machining of ceramic materials is calculated at higher closeness index (CI) values.

4.5 Summary

This chapter provides an integrated MCDM method comprising of fuzzy-MCRA methods for modeling and optimization of USM process. To check the effectiveness of the proposed technique, an experimental study has been conducted on optimization of USM process while machining zirconia (ZrO_2) composite. The result shows that experiment number 3 leads to the optimum output responses for USM process.

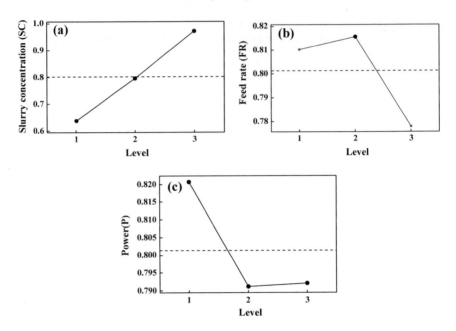

Fig. 4.4 a–c Response graphs of closeness index (CI) values

The combination of process parameters for the optimal setting is slurry concentration (SC) as 40%, feed rate (FR) as 1.4 mm/min, and power (P) as 400 W. The optimal setting provides better MRR with less taper angle and overcut during machining of zirconia (ZrO_2) composite by USM process. Additionally, the recommended optimal setting minimizes the manufacturing cost, production time, and wastage while improving the product quality and efficiency of the ultrasonic machining process. In essence, it is concluded that fuzzy-MCRA technique has potential to be effectively used for modeling and optimization of various advanced machining processes.

References

1. Thoe TB, Aspinwall DK, Wise MLH (1998) Review on ultrasonic machining. Int J Mach Tools Manuf 38(4):239–255
2. Benedict GF (1987) Non-traditional manufacturing processes. Marcel Dekker, New York
3. Gupta K, Jain NK, Laubscher RF (2016) Chapter-4 Assisted hybrid machining processes. In: Gupta K, Jain NK, Laubscher RF (eds) Hybrid machining processes. Springer, Cham
4. Kumar J (2013) Ultrasonic machining—a comprehensive review. Mach Sci Technol 17(3):325–379
5. Gill SS, Singh J (2010) An adaptive neuro-fuzzy inference system modeling for material removal rate in stationary ultrasonic drilling of sillimanite ceramic. Expert Syst Appl 37(8):5590–5598

6. Chakravorty R, Gauri SK, Chakraborty S (2013) Optimization of multiple responses of ultrasonic machining (USM) process: a comparative study. Int J Ind Eng Comput 4(2):285–296

7. Jain NK, Jain VK, Deb K (2007) Optimization of process parameters of mechanical type advanced machining processes using genetic algorithms. Int J Mach Tools Manuf 47(6):900–919

8. Goswami D, Chakraborty S (2015) Parametric optimization of ultrasonic machining process using gravitational search and fireworks algorithms. Ain Shams Eng J 6(1):315–331

9. Kumar J, Khamba JS (2010) Multi-response optimisation in ultrasonic machining of titanium using Taguchi's approach and utility concept. Int J Manuf Res 5(2):139–160

10. Kumar V, Khamba JS (2009) Parametric optimization of ultrasonic machining of co-based super alloy using the Taguchi multi-objective approach. Prod Eng Res Dev 3:417–425

11. Rao RV, Pawar PJ, Davim JP (2010) Parameter optimization of ultrasonic machining process using non-traditional optimization algorithms. Mater Manuf Process 25(10):1120–1130

12. Jagadish Ray A (2015) Green cutting fluid selection using multi-attribute decision making approach. J Inst Eng Ser C 96:35–39

13. Chakravorty R, Gauri SK, Chakraborty S (2013) Optimization of multiple responses of ultrasonic machining (USM) process: a comparative study. Int J Ind Eng Comput 4:285–296

14. Logothetis N, Haigh A (1998) Characterizing and optimizing multi-response processes by the taguchi method. Qual Reliab Eng Int 4:159–169

Chapter 5
Modeling and Optimization of Rapid Prototyping Process

5.1 Introduction

Rapid prototyping (RP) is an application of additive layer manufacturing (ALM) where complex engineered parts or assemblies are made from their CAD models by layered deposition of material [1]. The ALM technology was developed in 1980. Since then, there has been tremendous research and development to establish this field further, which resulted in commercialization of various ALM systems based on novel techniques such as stereolithography, selective laser sintering, fused deposition modeling, laminated object manufacturing, and 3D printing [2]. Basically, the ALM processes can majorly be classified as liquid-based, solid-based, and powder-based processes. A solid ALM process, i.e., fused deposition modeling (FDM) originated in 1989, is a major process used in rapid prototyping to produce parts and products for testing and evaluation purposes.

Energy- and resource-efficient production, excellent part quality, limited human errors, eliminating the procurement of material of specific shape and size, and eliminating detailed process planning, etc., are some of the benefits of ALM processes.

In FDM and other ALM processes, first, a computer-aided design (CAD) model is created and then converted to stereolithography (STL) format. Second, the converted stereolithography model is then fed to the rapid prototyping machine for slicing using various slicing techniques. Last, the sliced model is then fed to the rapid prototyping machine for development of model layer by layer until the finished product is obtained. Figure 5.1 presents the aforementioned steps.

In this chapter, modeling and optimization of FDM process using MCDM techniques is done. Before introducing the MCDM optimization techniques used in this work and discussing FDM-based experimental study followed by optimization, it is worth introducing the working principle of FDM technique, which is as follows.

In FDM, raw material (a thermoplastic filament or metallic wire) is heated to its melting point and then extruded layer by layer to build the part model [1, 2]. Figure 5.2 illustrates the working principle of FDM process. Extrusion nozzles, support and

© The Author(s), under exclusive license to Springer Nature Switzerland AG 2019
S. Bhowmik et al., *Modeling and Optimization of Advanced Manufacturing Processes*,
Manufacturing and Surface Engineering, https://doi.org/10.1007/978-3-030-00036-3_5

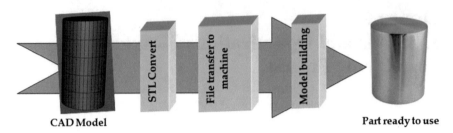

Fig. 5.1 Typical steps of product manufacturing in ALM

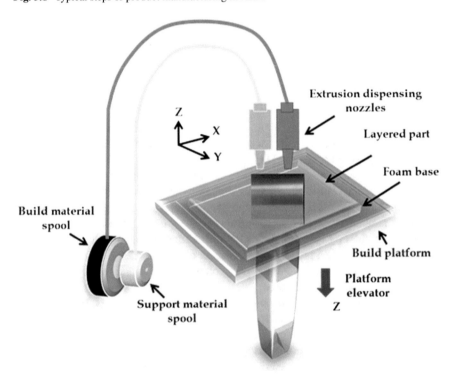

Fig. 5.2 Working principle of FDM process

build material spools, platform, and control system are the main parts of any FDM system/machine. The plastic filament or metallic wire is first unwounded from its spool and supplied to the extrusion nozzles where the material is melted and extruded onto the platform. The nozzle movement is controlled to follow CAD model-based CNC tool path in X-, Y-, and Z-directions. The layers built on the platform are cooled and hardened and kept extruded until the specified model geometry is formed [1, 2].

The quality and performance of the parts produced by FDM depend on its system parameters. The system parameters include both process/input parameters such as hatch spacing, layer thickness, scanning velocity, power, shell thickness, orientation angle, fill density, and output/response parameters such as dimensional accuracy, surface roughness, part strength, process time, and process energy. Improper selection of the process or input parameters for the rapid prototyping process may lead to major problems in manufactured part quality such as error in the dimensional accuracy, low strength, high energy consumption, long manufacturing time, low productivity, and poor finish and surface quality.

Further, the improper selection of the process parameters affects the process performance, efficiency, as well as cost of the product. Traditionally, the input or process parameters are selected based on operator experience and handbook values. However, the selected input or process parameters may not be optimal because of the operational errors occur due to judgments of the operators which lead to make more experimental runs and time-consuming. In spite of the above-mentioned issues related to selection of process parameters for rapid prototyping process, the rapid prototyping process also includes diversified response parameters; i.e., some of the parameters are of higher the better while other of lower the better which are based on beneficial and non-beneficial criteria. For example, part strength and dimensional accuracy are considered as beneficial criteria in which higher values are considered for better performance in rapid prototyping process. Moreover, process time and surface roughness are considered as non-beneficial criteria in which lower values are considered for better performance. This leads to selecting the optimal process parameters for rapid prototyping process a more complex and challenging task. Hence, modeling and optimization of rapid prototyping process is essential. In addition, selecting the optimal process parameters for rapid prototyping process is considered to be multi-criteria decision-making (MCDM) problem. The rapid prototyping process involves many input process parameters (i.e., alternatives) and output parameters (i.e., criteria).

In fact, many investigators have worked on modeling and optimization of process parameters in RP process such as bacterial foraging technique [3], response surface methodology [4],Taguchi method [5–7], finite element analysis (FEA) [8], generic algorithm [9], genetic algorithm and FEA virtual simulation [10], full factorial design [11], fuzzy and artificial neural network [12], grey relational analysis, and artificial neural network [13], and Teaching–learning-based algorithm [14]. It is seen that characteristics of the parts fabricated by RP processes are the functions of various input/process parameters. The process can be significantly improved by proper adjustment of the process parameters. Moreover, existing methods do not consider the uncertain, imprecise, and vagueness information of response parameters during the modeling and optimization of rapid prototyping process. In addition, existing methods failed to consider the effect of both interrelationships between the input/output parameters and beneficial and non-beneficial criteria separately during the modeling and optimization of rapid prototyping process which results in low output quality. Considering the drawbacks of these methods, in the present chapter, an attempt is made to verify the integrated MCDM method for modeling and optimizing the RP process.

5.2 Integrated Fuzzy–M-COPRAS-Based MCDM Method

In this section, an integrated MCDM method is discussed. The integrated MCDM method consists of fuzzy with M-COPRAS (modified-complex proportional assessment). Fuzzy method is used to define the importance (or priority weights) of each criterion or output parameters considering uncertain, imprecise, and vagueness information of parameters. While M-COPRAS is used for optimal alternatives or input parameters for the rapid prototyping process considering the effect of interrelationship between the input and output parameters as well as both beneficial and non-beneficial criteria separately. The detailed steps of this integrated MCDM method are as follows.

Step 1: Construction of the Decision Matrix

Optimization of the rapid prototyping process starts with the design of decision matrix. The decision matrix includes number of criteria/response parameters with their corresponding alternatives/process parameters. The decision matrix is developed based on design of experimental methods which depends on the number of process parameters and their levels. The decision matrix D_{ij} is developed using the following expression.

$$D_{ij} = \begin{bmatrix} & C_1 & C_2 & \cdots & C_n \\ A_1 & Y_{11} & Y_{12} & \dots & Y_{1j} \\ A_2 & Y_{21} & Y_{22} & \cdots & Y_{2j} \\ \vdots & \vdots & \vdots & \dots\dots & \vdots \\ A_i & Y_{i1} & Y_{i2} & \cdots & Y_{ij} \end{bmatrix} \quad \text{for } i = 1, 2, 3, 4 \dots m \text{ and } j = 1, 2, 3, 4 \dots n$$

(5.1)

where D_{ij} is the assessment value of ith alternative/input parameter on jth criteria/response parameters, A_1, A_2, A_3, ...A_i represent alternatives/process parameter/experimental setting of rapid prototyping process, C_1, C_2, C_3, ...C_j represent criteria/response parameters of the rapid prototyping process, and Y_{11}, Y_{12}, ...Y_{mn} represent performance values of the criteria/response parameters corresponding to each input parameter setting.

Step 2: Normalization of the Decision Matrix

Then, normalization of the decision matrix is carried out using Eq. (5.2). The normalization is done for making the different units of rapid prototyping process parameters into comparable units.

$$G_{ij} = \frac{D_{ij}}{\sum_{i=1}^{n} D_{ij}}$$

(5.2)

where G_{ij} represents the performance weights of ith alternatives/input parameters settings/experimental runs on jth criteria/output parameters, n represents the number of criteria/output parameters, and D_{ij} represents a dimensionless number in the range [0, 1] interval representing the normalized performance weights of ith alternatives/input parameters settings/experimental runs on jth criteria/output parameters.

Step 3: Determination of Priority Weights of the Response Parameters Using fuzzy Set Theory

After the normalization of the decision matrix, the weights of each criteria/output parameters according to their relative importance are calculated using fuzzy set theory. First, triangular fuzzy numbers are considered according to the importance expressed in linguistic variables. Based on the importance of each criterion/output parameter, four experts' opinions are taken with the help of triangular fuzzy numbers. Then, development of fuzzy decision matrix is carried out by expressing each of the each criterion/output parameter in triangular fuzzy numbers using the following expressions.

$$F = \begin{bmatrix} x11 & x12 & \cdots & x1n \\ x21 & x22 & \cdots & x2n \\ \vdots & \cdots & \cdots & \vdots \\ xm1 & xm2 & \cdots & xmn \end{bmatrix} \quad (\text{for } i = 1, 2, 3, 4 \ldots m \text{ and } j = 1, 2, 3, 4 \ldots n)$$

(5.3)

$$x_{ij} = \left(x_{ij1}, x_{ij2}, x_{ij3} \right)$$

where x_{ij} is the criteria/output parameters' weight by the decision maker expressed in triangular fuzzy number.

Second, the aggregated fuzzy weights are calculated for each of the criteria/output parameters from triangular fuzzy numbers. Finally, defuzzification of the aggregated fuzzy decision matrix is done to transform the fuzzy weights of the criteria/output parameters into crisp weights (real weights). The center of area method is used for best non-fuzzy performance of each criteria/output parameters which is expressed in fuzzy weights. Defuzzification of the aggregated fuzzy decision matrix is formulated using the following equation to derive crisp values of the each criterion/output parameter.

$$w_j = \frac{\left[\left(U x_{ij} - L x_{ij} \right) + \left(M x_{ij} - L x_{ij} \right) \right]}{3} + L x_{ij}$$

$$(\text{for } i = 1, 2, 3, 4 \ldots m \text{ and } j = 1, 2, 3, 4 \ldots n) \qquad (5.4)$$

where U, M, L represent higher, medium, and lower limitations of the fuzzy weights of the criteria/output parameters in the aggregated fuzzy decision matrix.

Step 4: Determination of Weighted Normalized Decision Matrix

In the present step, the normalized experimental data of the criteria/output parameters for rapid prototyping process is further multiplied with the respective priority weights of the criteria/output parameters obtained by fuzzy set theory. The weighted normalized decision matrix is determined by Eq. (5.5).

$$X_{ij}^* = G_{ij} \times w_j \quad \text{(for } i = 1, 2, 3, 4 \ldots m \text{ and } j = 1, 2, 3, 4 \ldots n) \tag{5.5}$$

where X_{ij}^* indicates weighted normalized decision matrix, G_{ij} is normalized decision matrix, and w_j represents priority weight for each of the criteria/output parameters.

Step 5: Evaluation of the Relative Significance and Quantitative Utility Degree of the Alternatives

In this step, the overall evaluation of criteria/output parameters (i.e., beneficial and non-beneficial response parameters) is carried out via relative significance values. The relative significance values for each of the criteria/output parameters are determined using the following expressions.

$$S_i = \sum_{j=1}^{k} X_{Bij}^* \quad \text{(for } i = 1, 2, 3, 4 \ldots m, \ j = 1, 2, 3, 4 \ldots n \text{ and } k = \text{no of beneficial criteria)}$$

$$\tag{5.6}$$

$$M_i = \sum_{j=1}^{l} X_{NBij}^*$$

$$\text{(for } i = 1, 2, 3, 4 \ldots m, \ j = 1, 2, 3, 4 \ldots n \text{ and } l = \text{no. of non - beneficial criteria)} \tag{5.7}$$

$$E_i = S_i + \frac{M_{\min} \sum_{i=1}^{n} M_i}{M_i \sum_{i=1}^{n} \frac{M_{\min}}{M_i}} \quad \text{(for } i = 1, 2, 3, 4 \ldots m) \tag{5.8}$$

where S_i and M_i represent the summation of the weighted normalized values of beneficial and non-beneficial criteria/output parameters; X_{Bij} and X_{NBij} denote the beneficial criteria/output parameters and non-beneficial criteria/output parameters terms in the weighted normalized decision matrix. E_i represents relative significance values of each alternatives/input parameter setting; M_{min} represents minimum response value in the weighted normalized values of non-beneficial criteria/output parameters.

Step 6: Ranking of the Alternatives

At last, ranking of alternatives/experimental settings or input parameter settings is done based on the relative significance values of each of the response parameters. The alternative/experimental setting with highest relative significance value (E_i) indicates the most optimal setting while with the least relative significance value (E_i) indicates least optimal setting among the other experimental settings.

5.3 Experimental Study

5.3.1 Background

An experimental study has been conducted to demonstrate the modeling and parameter optimization of fused deposition modeling (FDM) process using integrated MCDM method. Work material, i.e., nylon 618, and alternatives/input parameters, namely layer height (LT), layer thickness (ST), orientation angle (OA), and criteria/output parameters, namely ultimate tensile strength, dimensional accuracy, and process time, are considered for experimentation. The quality and other characteristics of the fabricated parts are dependent on the input parameters considered in the FDM process. Improper adjustment of the input/process parameters results in poor dimensional accuracy, lower strength, and time-consuming while manufacturing parts in fused deposition modeling process. To select the optimal alternative/experimental setting or the optimal input parameters among other input parameter settings, modeling and optimization of the RF process is essential.

In this regard, in the present chapter, an attempt is made to demonstrate the integrated MCDM method, i.e., fuzzy set theory with M-COPRAS for modeling and optimization of fused deposition modeling process. In this method, fuzzy set theory is used for extraction of precise priority weights of the criteria/output parameters considering the uncertainty, imprecise, and vagueness parameters. M-COPRAS method is employed for selecting the optimal alternative/input parameters considering the effect of both beneficial and non-beneficial criteria/output parameters in account.

5.3.2 Workpiece Specimen

The work specimen of nylon 618 [mixture of polyamide 6/(polycaprolactam) and polyamide 66, i.e., (mixture of hexamethylenediamine and adipic acid)] is used in the experimentation. The work material possesses excellent tensile strength of 31.54 MPa, ultimate elongation of 86%, and less weight compared with other filament materials like acrylonitrile butadiene styrene (ABS) and polylactic acid or polylactide (PLA). First, rectangular bar specimen of dimension $150 \times 10 \times 4$ mm as per ASTM D638 standard and cube of dimension $10 \times 10 \times 10$ mm are made (Figs. 5.3 and 5.4) using 3D modeling software. Second, developed 3D model is converted into stereolithography (STL) format and sliced using CURA software. At last, sliced model is processed to FDM machine for fabrication of the part.

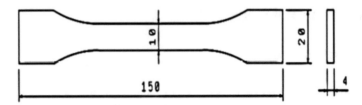

Fig. 5.3 Dimensions used for tensile specimen

Fig. 5.4 Cube specimen for dimensional accuracy

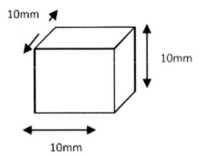

5.3.3 Experimental Procedure

The actual experimentation is carried out using actual setup on FDM machine (Fig. 5.5a) (pramaan Mini) manufactured by martinjn elserman. During experimentation, the voltage of 100–240 V and current of 4 A are supplied to the FDM machine.

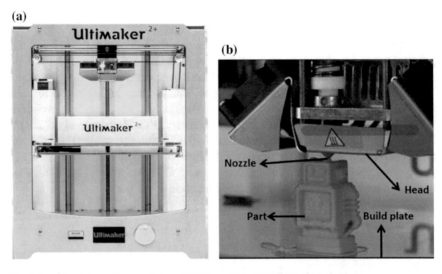

Fig. 5.5 **a** Fused deposition modeling (FDM) machine. **b** FDM tool head

Table 5.1 Process parameters and their levels for FDM process

Process parameters	Symbol	Units	Level I	Level II	Level III
Shell thickness	ST	mm	0.4	0.8	1.2
Layer thickness	LT	mm	0.1	0.2	0.3
Orientation angle	OA	Degree	0	15	30

Table 5.2 Parameter combinations and experimental results in FDM study

Exp. No	Input parameters			Output parameters		
	ST (mm)	LT (mm)	OA (°)	Ultimate tensile strength (UTS) in (MPa)	Dimensional accuracy (DA) in (mm^3)	Manufacturing time (MT) in (min)
1	0.4	0.1	0	19.05	1018.66	60
2	0.8	0.1	15	19.18	1002.56	79
3	1.2	0.1	30	25.48	1041.85	101
4	0.8	0.2	0	9.51	1089.52	34
5	1.2	0.2	15	15.68	1080.74	42
6	0.4	0.2	30	7.71	1073.04	48
7	1.2	0.3	0	8.35	1053.34	25
8	0.4	0.3	15	8.83	1167.63	26
9	0.8	0.3	30	11.82	1051.79	34

The FDM tool head contains two nozzles: one for building of the prototype and other for building of support structure as shown in Fig. 5.5b. Filament with 1.75 mm diameter is used for fabrication of the part layer by layer. Layer height, shell thickness, and orientation angle are considered as input parameters with three levels each (Table 5.1), and ultimate tensile strength, dimensional accuracy, process time are output parameters (Table 5.2).

Based on the Taguchi (L9) orthogonal array, nine experiments have been conducted to manufacture the part by FDM. Simultaneously, response parameters like ultimate tensile strength (UTS), dimensional accuracy (DA), process time (PT) are determined for each of the experimental settings. Similarly, universal testing machine (UTM) with the capacity of 20 tons is used for determination of ultimate tensile strength for each of the nine specimens. Also, the dimensional accuracy is determined by digital caliper by measuring specimen dimensions in the directions of X-, Y-, and Z-axes. The experimental results of FDM process are shown in Table 5.2.

5.3.4 Modeling of FDM Process

Modeling of the FDM process is done using fuzzy-M-COPRAS method to validate the experimental results for getting the optimal setting or optimal input parameters. First, a decision matrix is developed using Taguchi (L9) orthogonal array (Table 5.2) based on the input parameters and their levels as shown in Table 5.1. The decision matrix contains number of criteria as output parameters and corresponding alternatives as input parameter setting or experimental runs. The number of criteria considered as ultimate tensile strength, dimensional accuracy, and process time while number of alternatives or input parameter setting as layer thickness (LT), orientation angle (OA), and shell thickness (ST). Second, normalization of the decision matrix is done to convert the different measurement data of the criteria or output parameters of the FDM process into comparable data using Eq. (5.2). The results of the normalization are shown in Table 5.3.

Third, the precise priority weights of the criteria or output parameters are determined using fuzzy set theory. In this method, first, each criterion or output parameter is assigned a linguistic variable by four different experts based on their importance

Table 5.3 Normalized decision matrix

Exp. No	Ultimate tensile strength (UTS)	Dimensional accuracy (DA)	Manufacturing time (MT)
1	0.7476	0.8724	0.5941
2	0.7527	0.8586	0.7822
3	1	0.8923	1
4	0.3732	0.9331	0.3366
5	0.6154	0.9256	0.4158
6	0.3026	0.919	0.4752
7	0.3277	0.9021	0.2475
8	0.3465	1	0.2574
9	0.4639	0.9008	0.3366

Table 5.4 Linguistic variables expressed with triangular fuzzy numbers

Sl. No	Importance and abbreviation	Fuzzy numbers
1	Extremely low (EL)	(0, 0, 1)
2	Very low (VL)	(0, 0.1, 0.3)
3	Low (L)	(0.1, 0.3, 0.5)
4	Medium (M)	(0.3, 0.5, 0.7)
5	High (H)	(0.5, 0.7, 0.9)
6	Very high (VH)	(0.7, 0.9, 1)
7	Extremely high (EH)	(0.9, 1, 1)

Table 5.5 Linguistic variables assigned by decision makers

Response parameters	DM1	DM2	DM3	DM4
Ultimate tensile strength	VH	EH	VH	H
Dimensional accuracy	EH	H	VH	H
Manufacturing time	EL	VL	L	VL

Table 5.6 Aggregated fuzzy decision matrix

Output responses	Fuzzy weight-1	Fuzzy weight-2	Fuzzy weight-3
Ultimate tensile strength	0.7000	0.8750	0.9750
Dimensional accuracy	0.6500	0.8250	0.9500
Manufacturing time	0.0250	0.1250	0.3000

Table 5.7 Defuzzified aggregated decision matrix

Output Responses	Crisp weight
Ultimate tensile strength (UTS)	0.85
Dimensional accuracy (DA)	0.8083
Manufacturing time (MT)	0.15

as shown in Table 5.5. The assigned linguistic variables are then converted into their corresponding triangular fuzzy numbers using Table 5.4 and averaged to obtain the fuzzy weights of the criteria or output parameters as depicted in Table 5.6.

Next, the fuzzy weights are converted into crisp weights (real weights) by defuzzification using Eq. (5.4). This step is carried out for converting the three priority weights of each criterion or output parameters obtained in the previous step into single weight for ease of calculation. The defuzzified criteria weights are shown in Table 5.7. It is observed that ultimate tensile strength (UTS) yields highest crisp weight with 0.8500 and manufacturing time (MT) is found to be the least with 0.1500.

In this step, the crisp weight of each criterion is multiplied with the corresponding alternative values of normalized decision matrix (Table 5.3). The weighted normalized decision matrix is shown in Table 5.8.

The final step of the methodology is the calculation of relative significance values. The relative significance values are calculated using Eq. (5.8) which consists of S_i (sum of beneficiary criteria values) and M_i (sum of non-beneficiary criteria values) terms. In the present chapter, criteria, i.e., ultimate tensile strength (UTS) and dimensional accuracy (DA), are identified as beneficiary criteria/output parameters (higher the better) and manufacturing time (MT) as non-beneficiary criteria/output parameters (lower the better). The S_i (sum of beneficiary criteria values) and M_i (sum of non-beneficiary criteria values) are calculated using Eqs. (5.6) and (5.7), and results are shown in Table 5.9. Similarly, using the S_i and M_i values, the relative significance values (E_i) of each alternative are calculated which is shown in Table 5.9.

Table 5.8 Weighted normalized decision matrix

Exp. No	Ultimate tensile strength (UTS)	Dimensional accuracy (DA)	Manufacturing time (MT)
1	0.6355	0.7052	0.0891
2	0.6398	0.6947	0.1173
3	0.85	0.7213	0.15
4	0.3172	0.7543	0.0505
5	0.5281	0.7482	0.0624
6	0.2572	0.7429	0.0713
7	0.2786	0.7292	0.0371
8	0.2946	0.8083	0.0386
9	0.3943	0.7281	0.0505

Table 5.9 Maximum (S_i) and minimum (M_i) and relative significance (E_i) values

Exp. No	Sum of beneficial terms (S_i)	Sum of non-beneficial terms (M_i)	Relative significances (E_i)
1	1.3407	0.0891	1.391
2	1.3339	0.1173	1.3721
3	1.5713	0.15	1.6011
4	1.0715	0.0505	1.1602
5	1.2713	0.0624	1.3430
6	1.0001	0.0713	1.0629
7	1.0078	0.0371	1.1284
8	1.1029	0.0386	1.2189
9	1.1224	0.0505	1.2111

In the last step, ranking of the alternatives/input parameters using the relative significance values is done. The alternatives/input parameters with the highest relative significance value give the most optimum parameters. The experiment number 3 yields highest relative significance value of 1.6011, and hence, the corresponding parameter settings are the most optimum input parameters.

5.4   Result and Discussions

5.4.1   Optimum Combination of Process Parameters

The optimization of the FDM process parameters is done based on the relative significance values using integrated MCDM method. The relative significance values for each of the FDM criteria/output parameters are determined using the proposed method, and the results are in Table 5.10.

Table 5.10 Ranking of the FDM alternatives of FDM process

Exp. No	Input parameters			Output parameters			Relative significance (E_i) values	Rank
	ST (mm)	LT (mm)	OA (degree)	UTS (MPa)	DA (mm³)	MT (min)		
1	0.4	0.1	0	19.05	1018.66	60	1.3910	2
2	0.8	0.1	15	19.18	1002.56	79	1.3721	3
3	1.2	0.1	30	25.48	1041.85	101	1.6011	1
4	0.8	0.2	0	9.51	1089.52	34	1.1602	7
5	1.2	0.2	15	15.68	1080.74	42	1.3430	4
6	0.4	0.2	30	7.71	1073.04	48	1.0629	9
7	1.2	0.3	0	8.35	1053.34	25	1.1284	8
8	0.4	0.3	15	8.83	1167.63	26	1.2189	5
9	0.8	0.3	30	11.82	1051.79	34	1.2111	6

The result shows that experiment number 3 yields the highest relative significance value which indicates the optimal setting among the other alternatives or experimental runs or process parameter settings. The relative significance value for alternative or experimental run or process parameter setting is 1.6011 and is the highest among the other relative significance values. The optimal settings obtained are shell thickness (ST) as 1.2 mm, orientation angle (OA) as 30°, and layer thickness as 0.1 mm. The corresponding output parameters obtained are ultimate tensile strength as 25.48 MPa, dimensional accuracy as 1041.85 mm³, and manufacturing time as 101 min. The optimal process parameters of FDM process provides higher ultimate tensile strength, good dimensional accuracy, and lesser process time as well as improves the productivity and efficiency of the FDM process.

5.4.2   Optimum Input Parameters and Level Combination

In this section, determination of optimum input parameters and their level combination has been illustrated. The optimum input parameters are determined by averaging the relative significance (E_i) values for each level of input parameters, and results are shown in Table 5.11 The result shows that relative significance (E_i) values for each of the input parameters such as ultimate tensile strength (UTS), dimensional accuracy (DA), and manufacturing time (MT) show better correlation between the reference and comparability sequence. Hence, comparability sequence has relative significance (E_i) values for optimum parameters which are shell thickness (ST) at level 3, orientation angle (OA) at level 2, and layer thickness (LT) at level 2.

Furthermore, most significant input parameters are determined by calculating the difference between maximum and minimum values of the mean relative significance (E_i) values. The mean relative significance (E_i) values for each of the input parameters

Table 5.11 Average relative significance (E_i) values

Input parameters	Relative significance (E_i) values			Max − Min	Rank
	Level 1	Level 2	Level 3		
ST	1.224	1.248	1.358[a]	0.133	2
LT	1.455[a]	1.189	1.186	0.269	1
OA	1.227	1.311[a]	1.292	0.085	3

[a]Optimal level

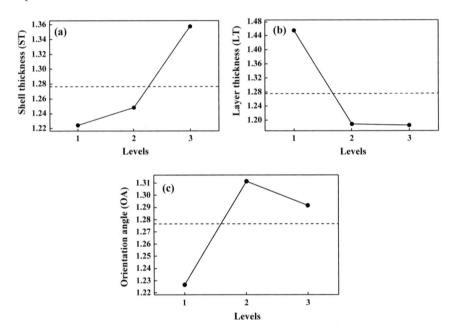

Fig. 5.6 a–c Response graphs of relative significance (E_i) values

obtained are shell thickness (ST) as 0.133, orientation angle (OA) as 0.085, and layer thickness (LT) as 0.269. It is observed from the results that parameter layer thickness (LT) is found to be more significant input parameter while orientation angle (OA) as least significant parameter for FDM process. The order of importance of the input parameters for FDM process can be listed as layer thickness (LT)—shell thickness (ST)—orientation angle (OA).

Additionally, the influence of each input parameter can be seen in the graphs as shown in Fig. 5.6. This shows that small changes in the relative significance (E_i) values when a factor goes from their level 1–3 and greater relative significance (E_i) values provide the optimum values for the FDM process.

5.5 Summary

Modeling and optimization of rapid prototyping process using an integrated multi-criteria decision-making (MCDM) method is discussed in the present chapter. It consists of fuzzy technique coupled with M-COPRAS methods. An experimental case study on modeling and optimization of fused deposition modeling (FDM) is presented using the integrated MCDM method.

The optimization result shows that the experiment number 3 yields highest relative significance values among the other optimal settings of the FDM process. The optimal process parameters obtained are shell thickness (ST) as 1.2 mm, orientation angle (OA) as 30°, and layer thickness as 0.1 mm. The optimal parameter of FDM process provides most optimal response parameters such as higher ultimate tensile strength, better dimensional accuracy, and lesser process time. This subsequently increases the quality and productivity as well as improves the performance and efficiency of the FDM process. Thus, it is concluded that the hybrid technique, i.e., fuzzy with M-COPRAS method, can be used as systematic approach for modeling and optimization of any other advanced manufacturing specially rapid prototyping processes.

References

1. Gibson I, Rosen D, Stucker B (2015) Additive manufacturing technologies: 3D printing, rapid prototyping, and direct digital manufacturing, 2nd edn. Springer Science and Business Media Pvt Ltd, New York
2. Gupta K, Jain NK, Laubscher RF (2017) Chapter 4: Advances in gear manufacturing. In: Advanced gear manufacturing and finishing-classical and modern processes. Academic Press Inc
3. Panda SK, Padhee S, Anoop Kumar S, Mahapatra SS (2009) Optimization of fused deposition modeling (FDM) process parameters using bacterial foraging technique. J Intell Info Manag 1:89–97
4. Mohamed OA, Masood SH, Bhowmik JL (2016) Mathematical modeling and FDM process parameters optimization using response surface methodology based on Q-optimal design. J Appl Math Model 40:10052–10073
5. Kumar N, Kumar H, Khurmi JS (2016) Experimental Investigation of process parameters for rapid prototyping technique (Selective Laser Sintering) to enhance the part quality of prototype by Taguchi method (vol 23). In: 3rd ICIAME Procedia Technology, pp 352–360
6. Raju BS, Shekar UC, Venkateswarlu K, Drakashayani DN (2014) Establishment of Process model for rapid prototyping technique (stereolithography) to enhance the part quality by Taguchi method (vol 14). In: 2nd ICIAME 2014 Procedia Technology, pp 380–389
7. Phatak AM, Pandee SS (2012) Optimum part orientation in rapid prototyping using generic algorithm. J Manuf Syst 31:395–402
8. Alafaghani A, Qattawi A, Alrawi B, Guzman A (2017) Experimental optimization of fused deposition modelling processing parameters: a design for manufacturing approach (vol 10). In: 45th SME NAMRC conference. Procedia Manufacturing, pp 791–803
9. Lee BH, Abdullah J, Khan ZA (2005) Optimization of rapid prototyping parameters for production of flexible ABS object‖. J Mater Process Technol 169:54–61
10. Villalpando L, Eiliat H, Urbanic RJ(2014) An optimization approach for components built by fused deposition modeling with parametric internal structures, product services systems

and value creation. In: Proceedings of the 6th CIRP conference on industrial, product-service systems, pp 800–805

11. Kumar S, Kannan VN, Sankaranarayanan G (2014) Parameter optimization of ABS-M30i parts produced by fused deposition modeling for minimum surface roughness. Int J Curr Eng Technol 3:93–97

12. Nidagundi VB, Keshavamurthy R, Prakash (2015) CPS studies on parametric optimization for fused deposition modelling process. In: 4th International conference on materials processing and characterization proceedings (vol 2), pp 1691–1699

13. Anoop Kumar S, Ohdar RK Siba Sankar Mahapatra (2009) Improving dimensional accuracy of fused deposition modelling processed part using grey Taguchi method. In: 4th International conference on materials processing and characterization proceedings, vol 30, issue no 10, pp 4243–4252

14. Rao PV, Rai DP(2016) Optimization of fused deposition modeling process using teaching learning based algorihm. Int J Eng Sci Technol 587–603

Index

Printed in the United States
By Bookmasters